Applied Statistics and Probability for Engineers

Second Edition
Selected Topics

Douglas C. Montgomery
George C. Runger

prepared for use in the
Department of Industrial Engineering
Arizona State University
Temple, Arizona

WILEY
CUSTOM SERVICES

Printed in the United States of America.

ISBN 0-471-03265-4

Applied Statistics and Probability for Engineers

Second Edition

Douglas C. Montgomery
Arizona State University

George C. Runger
Arizona State University

John Wiley & Sons, Inc.
New York • Chichester • Brisbane • Toronto • Singapore

EDITOR *Wayne Anderson*
MARKETING MANAGER *Katherine Hepburn*
SENIOR PRODUCTION MANAGER *Lucille Buonocore*
SENIOR PRODUCTION EDITOR *Monique Calello*
SENIOR DESIGNER *Harry Nolan*
ILLUSTRATION COORDINATOR *Jaime Perea*
ILLUSTRATION STUDIO *Wellington Studios*
COVER DESIGN & ILLUSTRATION *Lynn Rogan*

This book was set in 10/12 Times Roman by Monotype Composition Company and printed and bound by Quebecor Printing, Fairfield. The cover was printed by Phoenix Color Corporation.

This book is printed on acid-free paper. ∞

The paper in this book was manufactured by a mill whose forest management programs include sustained yield harvesting of its timberlands. Sustained yield harvesting principles ensure that the number of trees cut each year does not exceed the amount of new growth.

Library of Congress Cataloging-in-Publication Data

Montgomery, Douglas C.
Applied statistics and probability for engineers / Douglas C. Montgomery, George C. Runger.—2nd ed.
p. cm.
Includes index.
ISBN 0-471-17027-5 (acid-free paper)
1. Statistics. 2. Probabilities. I. Runger, George C. II. Title.
QA276.12.M645 1999
519.5—dc21 98-34994
CIP

Printed in the United States of America

10 9 8 7 6 5 4 3 2 1

Contents

CHAPTER 11 Multiple Linear Regression 483

APPENDICES 819

INDEX I-1

8 Statistical Inference for a Single Sample

CHAPTER OUTLINE

8-6 SUMMARY TABLE OF INFERENCE PROCEDURES FOR A SINGLE SAMPLE

The table on the end papers of this book (inside front cover) presents a summary of all the single-sample inference procedures from this chapter. The table contains the null hypothesis statement, the test statistic, the various alternative hypotheses and the criteria for rejecting H_0, and the formulas for constructing the $100(1 - \alpha)\%$ confidence interval.

8-7 TESTING FOR GOODNESS OF FIT

The hypothesis-testing procedures that we have discussed in previous sections are designed for problems in which the population or probability distribution is known and the hypotheses involve the parameters of the distribution. Another kind of hypothesis is often encountered: we do not know the underlying distribution of the population, and we wish to test the hypothesis that a particular distribution will be satisfactory as a population model. For example, we might wish to test the hypothesis that the population is normal.

In Chapter 5, we discussed a very useful graphical technique for this problem called **probability plotting** and illustrated how it was applied on the case of a normal distribution. In this section, we describe a formal goodness-of-fit test procedure based on the chi-square distribution.

The test procedure requires a random sample of size n from the population whose probability distribution is unknown. These n observations are arranged in a frequency histogram, having k bins or class intervals. Let O_i be the observed frequency in the ith class interval. From the hypothesized probability distribution, we compute the expected frequency in the ith class interval, denoted E_i. The test statistic is

$$X_0^2 = \sum_{i=1}^{k} \frac{(O_i - E_i)^2}{E_i} \qquad (8\text{-}81)$$

It can be shown that, if the population follows the hypothesized distribution, X_0^2 has, approximately, a chi-square distribution with $k - p - 1$ degrees of freedom, where p represents the number of parameters of the hypothesized distribution estimated by sample statistics. This approximation improves as n increases. We would reject the hypothesis that the distribution of the population is the hypothesized distribution if the calculated value of the test statistic $\chi_0^2 > \chi_{\alpha, k-p-1}^2$.

One point to be noted in the application of this test procedure concerns the magnitude of the expected frequencies. If these expected frequencies are too small, then the test statistic X_0^2 will not reflect the departure of observed from expected, but only the small magnitude of the expected frequencies. There is no general agreement regarding the minimum value of expected frequencies, but values of 3, 4, and 5 are widely used as minimal. Some writers suggest that an expected frequency could be as small as 1 or 2, so long as most of them exceed 5. Should an expected frequency be too small, it can be combined with the expected frequency in an adjacent class interval. The corresponding observed frequencies would then also be combined, and k would be reduced by one. Class intervals are not required to be of equal width.

We now give two examples of the test procedure.

EXAMPLE 8-18

A Poisson Distribution

The number of defects in printed circuit boards is hypothesized to follow a Poisson distribution. A random sample of $n = 60$ printed boards has been collected, and the number of defects observed. The following data result:

Number of Defects	Observed Frequency
0	32
1	15
2	9
3	4

The mean of the assumed Poisson distribution in this example is unknown and must be estimated from the sample data. The estimate of the mean number of defects per board is the sample average, that is, $(32 \cdot 0 + 15 \cdot 1 + 9 \cdot 2 + 4 \cdot 3)/60 = 0.75$. From the Poisson distribution with parameter 0.75, we may compute p_i, the theoretical, hypothesized probability associated with the ith class interval. Since each class interval corresponds to a particular number of defects, we may find the p_i as follows:

$$p_1 = P(X = 0) = \frac{e^{-0.75}(0.75)^0}{0!} = 0.472$$

$$p_2 = P(X = 1) = \frac{e^{-0.75}(0.75)^1}{1!} = 0.354$$

$$p_3 = P(X = 2) = \frac{e^{-0.75}(0.75)^2}{2!} = 0.133$$

$$p_4 = P(X \geq 3) = 1 - (p_1 + p_2 + p_3) = 0.041$$

The expected frequencies are computed by multiplying the sample size $n = 60$ times the probabilities p_i. That is, $E_i = np_i$. The expected frequencies are shown below.

Number of Defects	Probability	Expected Frequency
0	0.472	28.32
1	0.354	21.24
2	0.133	7.98
3 (or more)	0.041	2.46

Since the expected frequency in the last cell is less than 3, we combine the last two cells:

Number of Defects	Observed Frequency	Expected Frequency
0	32	28.32
1	15	21.24
2 (or more)	13	10.44

The chi-square test statistic in equation 8-81 will have $k - p - 1 = 3 - 1 - 1 = 1$ degree of freedom, because the mean of the Poisson distribution was estimated from the data.

The eight-step hypothesis-testing procedure may now be applied, using $\alpha = 0.05$, as follows:

1. The variable of interest is the form of the distribution of defects in printed circuit boards.
2. H_0: The form of the distribution of defects is Poisson.
3. H_1: The form of the distribution of defects is not Poisson.
4. $\alpha = 0.05$
5. The test statistic is

$$\chi_0^2 = \sum_{i=1}^{k} \frac{(o_i - E_i)^2}{E_i}$$

6. Reject H_0 if $\chi_0^2 > \chi_{0.05,1}^2 = 3.84$.

7. Computations:

$$\chi_0^2 = \frac{(32 - 28.32)^2}{28.32} + \frac{(15 - 21.24)^2}{21.24} + \frac{(13 - 10.44)^2}{10.44} = 2.94$$

8. Conclusions: Since $\chi_0^2 = 2.94 < \chi_{0.05,1}^2 = 3.84$, we are unable to reject the null hypothesis that the distribution of defects in printed circuit boards is Poisson. The P-value for the test is $P = 0.0864$. (This value was computed using an HP-48 calculator.)

EXAMPLE 8-19

A Continuous Distribution

A manufacturing engineer is testing a power supply used in a notebook computer. Using $\alpha = 0.05$, he wishes to determine whether output voltage is adequately described by a normal distribution. From a random sample of $n = 100$ units, he obtains sample estimates of the mean and standard deviation as $\bar{x} = 5.04$ V and $s = 0.08$ V.

A common practice in constructing the class intervals for the frequency distribution used in the chi-square goodness-of-fit test is to choose the cell boundaries so that the expected frequencies $E_i = np_i$ are equal for all cells. To use this method, we want to choose the cell boundaries $a_0, a_1, \ldots, a_k$ for the k cells so that all the probabilities

$$p_i = P(a_{i-1} \leq X \leq a_i) = \int_{a_{i-1}}^{a_i} f(x)\, dx$$

are equal. Suppose we decide to use $k = 8$ cells. For the standard normal distribution, the intervals that divide the scale into eight equally likely segments are [0, 0.32), [0.32, 0.675) [0.675, 1.15), [1.15, ∞) and their four "mirror image" intervals on the other side of zero. For each interval $p_i = 1/8 = 0.125$, so the expected cell frequencies are $E_i = np_i = 100(0.125) = 12.5$. The complete table of observed and expected frequencies is as follows.

Class Interval	Observed Frequency o_i	Expected Frequency E_i
$x < 4.948$	12	12.5
$4.948 \leq x < 4.986$	14	12.5
$4.986 \leq x < 5.014$	12	12.5
$5.014 \leq x < 5.040$	13	12.5
$5.040 \leq x < 5.066$	12	12.5
$5.066 \leq x < 5.094$	11	12.5
$5.094 \leq x < 5.132$	12	12.5
$5.132 \leq x$	14	12.5
Totals	100	100

The boundary of the first class interval is $\bar{x} - 1.15s = 4.948$. The second class interval is $[\bar{x} - 1.15s, \bar{x} - 0.675s)$ and so forth. We may apply the eight-step hypothesis-testing procedure to this problem.

1. The variable of interest is the form of the distribution of power supply voltage.
2. H_0: The form of the distribution is normal.
3. H_1: The form of the distribution is nonnormal.
4. $\alpha = 0.05$
5. The test statistic is

$$\chi_0^2 = \sum_{i=1}^{k} \frac{(o_i - E_i)^2}{E_i}$$

6. Since two parameters in the normal distribution have been estimated, the chi-square statistic above will have $k - p - 1 = 8 - 2 - 1 = 5$ degrees of freedom. Therefore, we will reject H_0 if $\chi_0^2 > \chi_{0.05,5}^2 = 11.07$.
7. Computations:

$$\begin{aligned}\chi_0^2 &= \sum_{i=1}^{8} \frac{(o_i - E_i)^2}{E_i} \\ &= \frac{(12 - 12.5)^2}{12.5} + \frac{(14 - 12.5)^2}{12.5} + \cdots + \frac{(14 - 12.5)^2}{12.5} \\ &= 0.64\end{aligned}$$

8. Conclusions: Since $\chi_0^2 = 0.64 < \chi_{0.05,5}^2 = 11.07$, we are unable to reject H_0, and there is no strong evidence to indicate that output voltage is not normally distributed. The P-value for the chi-square statistic $\chi_0^2 = 0.64$ is $P = 0.9861$.

EXERCISES FOR SECTION 8-7

8-58. Consider the following frequency table of observations on the random variable X.

Values	0	1	2	3	4
Observed Frequency	24	30	31	11	4

(a) Based on these 100 observations, is a Poisson distribution with a mean of 1.2 an appropriate model? Perform a goodness-of-fit procedure with $\alpha = 0.05$.

(b) Calculate the P-value for this test.

8-59. Let X denote the number of flaws observed on a large coil of galvanized steel. Seventy-five coils are inspected and the following data were observed for the values of X.

Values	1	2	3	4	5	6	7	8
Observed Frequency	1	11	8	13	11	12	10	9

(a) Does the assumption of Poisson distribution seem appropriate as a probability model for this data? Use $\alpha = 0.01$.

(b) Calculate the P-value for this test.

8-60. The number of calls arriving to a switchboard from noon to 1 PM during the business days Monday through Friday is monitored for 6 weeks, i.e., 30 days. Let X be defined as the number of calls during that one-hour period. The relative frequency of calls was recorded and reported as

Value	5	6	8	9	10
Relative Frequency	0.067	0.067	0.100	0.133	0.200

Value	11	12	13	14	15
Relative Frequency	0.133	0.133	0.067	0.033	0.067

(a) Does the assumption of a Poisson distribution seem appropriate as a probability model for this data? Use $\alpha = 0.05$.

(b) Calculate the P-value for this test.

8-61. Consider the following frequency table of observations on the random variable X.

Values	0	1	2	3	4
Frequency	4	21	10	13	2

(a) Based on these 50 observations, is a binomial distribution with $n = 6$ and $p = 0.25$ an appropriate model? Perform a goodness-of-fit procedure with $\alpha = 0.05$.

(b) Calculate the P-value for this test.

8-62. Define X as the number of underfilled bottles in a filling operation in a carton of 24 bottles. Sixty cartons are inspected and the following observations on X are recorded.

Values	0	1	2	3
Frequency	39	23	12	1

(a) Based on these 75 observations, is a binomial distribution an appropriate model? Perform a goodness-of-fit procedure with $\alpha = 0.05$.

(b) Calculate the P-value for this test.

8-63. The number of cars passing eastbound through the intersection of Mill Avenue and University Avenue has been tabulated by a group of civil engineering students. They have obtained the following data:

Vehicles per Minute	Observed Frequency	Vehicles per Minute	Observed Frequency
40	14	53	102
41	24	54	96
42	57	55	90
43	111	56	81
44	194	57	73
45	256	58	64
46	296	59	61
47	378	60	59
48	250	61	50
49	185	62	42
50	171	63	29
51	150	64	18
52	110	65	15

(a) Does the assumption of a Poisson distribution seem appropriate as a probability model for this process? Use $\alpha = 0.05$.

(b) Calculate the P-value for this test.

8-8 CONTINGENCY TABLE TESTS

Many times, the n elements of a sample from a population may be classified according to two different criteria. It is then of interest to know whether the two methods of classification are statistically independent; for example, we may consider the population of graduating engineers, and we may wish to determine whether starting salary is independent of academic disciplines. Assume that the first method of classification has r levels and that the second method has c levels. We will let O_{ij} be the observed frequency for level i of the first classification method and level j on the second classification method. The data would, in general, appear as shown in Table 8-3. Such a table is usually called an **$r \times c$ contingency table.**

We are interested in testing the hypothesis that the row-and-column methods of classification are independent. If we reject this hypothesis, we conclude there is some interaction between the two criteria of classification. The exact test procedures are difficult to obtain, but an approximate test statistic is valid for large n. Let p_{ij} be the probability that a randomly selected element falls in the ijth cell, given that the two classifications are independent. Then $p_{ij} = u_i v_j$, where u_i is the probability that a randomly selected element falls in row class i and v_j is the probability that a randomly selected element falls in column class j. Now, assuming independence, the estimators of u_i and v_i are

$$\hat{u}_i = \frac{1}{n}\sum_{j=1}^{c} O_{ij}$$

$$\hat{v}_j = \frac{1}{n}\sum_{i=1}^{r} O_{ij} \tag{8-82}$$

Therefore, the expected frequency of each cell is

$$E_{ij} = n\hat{u}_i\hat{v}_j = \frac{1}{n}\sum_{j=1}^{c} O_{ij}\sum_{i=1}^{r} O_{ij} \tag{8-83}$$

Then, for large n, the statistic

$$\chi_0^2 = \sum_{i=1}^{r}\sum_{j=1}^{c}\frac{(O_{ij} - E_{ij})^2}{E_{ij}} \tag{8-84}$$

has an approximate chi-square distribution with $(r - 1)(c - 1)$ degrees of freedom if the null hypothesis is true. Therefore, we would reject the hypothesis of independence if the observed value of the test statistic χ_0^2 exceeded $\chi^2_{\alpha,(r-1)(c-1)}$.

Table 8-3 An $r \times c$ Contingency Table

		Columns			
		1	2	...	c
Rows	1	O_{11}	O_{12}	...	O_{1c}
	2	O_{21}	O_{22}	...	O_{2c}
	⋮	⋮	⋮	⋮	⋮
	r	O_{r1}	O_{r2}	...	O_{rc}

Table 8-4 Observed Data for Example 8-20

Job Classification	Pension Plan 1	2	3	Totals
Salaried workers	160	140	40	340
Hourly workers	40	60	60	160
Totals	200	200	100	500

EXAMPLE 8-20

A company has to choose among three pension plans. Management wishes to know whether the preference for plans is independent of job classification and wants to use $\alpha = 0.05$. The opinions of a random sample of 500 employees are shown in Table 8-4.

To find the expected frequencies, we must first compute $\hat{u}_1 = (340/500) = 0.68$, $\hat{u}_2 = (160/500) = 0.32$, $\hat{v}_1 = (200/500) = 0.40$, $\hat{v}_2 = (200/500) = 0.40$, and $\hat{v}_3 = (100/500) = 0.20$. The expected frequencies may now be computed from equation 8-83. For example, the expected number of salaried workers favoring pension plan 1 is

$$E_{11} = n\hat{u}_1\hat{v}_1 = 500(0.68)(0.40) = 136$$

The expected frequencies are shown in Table 8-5.

The eight-step hypothesis-testing procedure may now be applied to this problem.

1. The variable of interest is employee preference among pension plans.
2. H_0: Preference is independent of salaried versus hourly job classification.
3. H_1: Preference is not independent of salaried versus hourly job classification.
4. $\alpha = 0.05$
5. The test statistic is

$$\chi_0^2 = \sum_{i=1}^{r} \sum_{j=1}^{c} \frac{(o_{ij} - E_{ij})^2}{E_{ij}}$$

6. Since $r = 2$ and $c = 3$, the degrees of freedom for chi-square are $(r - 1)(c - 1) = (1)(2) = 2$, and we would reject H_0 if $\chi_0^2 > \chi_{0.05,2}^2 = 5.99$.

Table 8-5 Expected Frequencies for Example 8-20

Job Classification	Pension Plan 1	2	3	Totals
Salaried workers	136	136	68	340
Hourly workers	64	64	32	160
Totals	200	200	100	500

7. Computations:

$$\chi_0^2 = \sum_{i=1}^{2}\sum_{j=1}^{3} \frac{(o_{ij} - E_{ij})^2}{E_{ij}}$$

$$= \frac{(160 - 136)^2}{136} + \frac{(140 - 136)^2}{136} + \frac{(40 - 68)^2}{68} + \frac{(40 - 64)^2}{64}$$

$$+ \frac{(60 - 64)^2}{64} + \frac{(60 - 32)^2}{32} = 49.63$$

8. Conclusions: Since $\chi_0^2 = 49.63 > \chi_{0.05,2}^2 = 5.99$, we reject the hypothesis of independence and conclude that the preference for pension plans is not independent of job classification. The P-value for $\chi_0^2 = 49.63$ is $P = 1.671 \times 10^{-11}$. (This value was computed from an HP-48 calculator.) Further analysis would be necessary to explore the nature of the association between these factors. It might be helpful to examine the table of observed minus expected frequencies.

Using the two-way contingency table to test independence between two variables of classification in a sample from a single population of interest is only one application of contingency table methods. Another common situation occurs when there are r populations of interest and each population is divided into the same c categories. A sample is then taken from the ith population, and the counts are entered in the appropriate columns of the ith row. In this situation we want to investigate whether or not the proportions in the c categories are the same for all populations. The null hypothesis in this problem states that the populations are **homogeneous** with respect to the categories. For example, when there are only two categories, such as success and failure, defective and nondefective, and so on, then the test for homogeneity is really a test of the equality of r binomial parameters. Calculation of expected frequencies, determination of degrees of freedom, and computation of the chi-square statistic for the test for homogeneity are identical to the test for independence.

EXERCISES FOR SECTION 8-8

8-64. A company operates four machines three shifts each day. From production records, the following data on the number of breakdowns are collected:

	Machines			
Shift	*A*	*B*	*C*	*D*
1	41	20	12	16
2	31	11	9	14
3	15	17	16	10

Test the hypothesis (using $\alpha = 0.05$) that breakdowns are independent of the shift. Find the P-value for this test.

8-65. Patients in a hospital are classified as surgical or medical. A record is kept of the number of times patients require nursing service during the night and whether or not these patients are on Medicare. The data are presented here:

Medicare	Patient Category	
	Surgical	Medical
Yes	46	52
No	36	43

Test the hypothesis (using $\alpha = 0.01$) that calls by surgical-medical patients are independent of whether the patients are receiving Medicare. Find the P-value for this test.

8-66. Grades in a statistics course and an operations research course taken simultaneously were as follows for a group of students.

Statistics Grade	Operation Research Grade			
	A	*B*	*C*	Other
A	25	6	17	13
B	17	16	15	6
C	18	4	18	10
Other	10	8	11	20

Are the grades in statistics and operations research related? Use $\alpha = 0.01$ in reaching your conclusion. What is the P-value for this test?

8-67. An experiment with artillery shells yields the following data on the characteristics of lateral deflections and ranges. Would you conclude that deflection and range are independent? Use $\alpha = 0.05$. What is the P-value for this test?

Range (yards)	Lateral Deflection		
	Left	Normal	Right
0–1,999	6	14	8
2,000–5,999	9	11	4
6,000–11,999	8	17	6

8-68. A study is being made of the failures of an electronic component. There are four types of failures possible and two mounting positions for the device. The following data have been taken:

Mounting Position	Failure Type			
	A	*B*	*C*	*D*
1	22	46	18	9
2	4	17	6	12

Would you conclude that the type of failure is independent of the mounting position? Use $\alpha = 0.01$. Find the P-value for this test.

8-69. A random sample of students is asked their opinions on a proposed core curriculum change. The results are as follows.

Class	Opinion	
	Favoring	Opposing
Freshman	120	80
Sophomore	70	130
Junior	60	70
Senior	40	60

Test the hypothesis that opinion on the change is independent of class standing. Use $\alpha = 0.05$. What is the P-value for this test?

10 Simple Linear Regression and Correlation

CHAPTER OUTLINE

10-1 EMPIRICAL MODELS

In many problems two or more variables are inherently related, and it is necessary to explore the nature of this relationship. **Regression analysis** is a statistical technique for modeling and investigating the relationship between two or more variables. For example, in a chemical process, suppose that the yield of the product is related to the process-operating temperature. Regression analysis can be used to build a model to predict yield at a given temperature level. This model can also be used for process optimization, such as finding the level of temperature that maximizes yield, or for process control purposes.

As an illustration, consider the data in Table 10-1. In this table y is the purity of oxygen produced in a chemical distillation process, and x is the percentage hydrocarbons that are present in the main condenser of the distillation unit. Figure 10-1 presents a **scatter diagram** of the data in Table 10-1. This is just a graph on which each (x_i, y_i) pair is represented as a point plotted in a two-dimensional coordinate system. Inspection of this scatter diagram indicates that, although no simple curve will pass exactly through all the points, there is a strong indication that the points lie scattered randomly around a

Table 10-1 Oxygen and Hydrocarbon Levels

Observation Number	Hydrocarbon Level x (%)	Purity y (%)
1	0.99	90.01
2	1.02	89.05
3	1.15	91.43
4	1.29	93.74
5	1.46	96.73
6	1.36	94.45
7	0.87	87.59
8	1.23	91.77
9	1.55	99.42
10	1.40	93.65
11	1.19	93.54
12	1.15	92.52
13	0.98	90.56
14	1.01	89.54
15	1.11	89.85
16	1.20	90.39
17	1.26	93.25
18	1.32	93.41
19	1.43	94.98
20	0.95	87.33

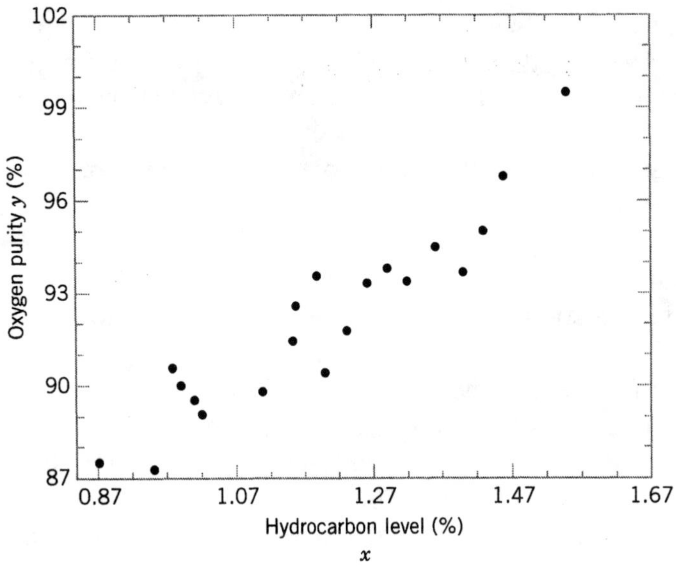

Figure 10-1 Scatter diagram of oxygen purity versus hydrocarbon level from Table 10-1.

straight line. Therefore, it is probably reasonable to assume that the mean of the random variable Y is related to x by the following straight-line relationship:

$$E(Y \mid x) = \mu_{Y|x} = \beta_0 + \beta_1 x$$

where the slope and intercept of the line are called **regression coefficients.** While the mean of Y is a linear function of x, the actual observed value y does not fall exactly on a straight line. The appropriate way to generalize this to a **probabilistic linear model** is to assume that the expected value of Y is a linear function of x, but that for a fixed value of x the actual value of Y is determined by the mean value function (the linear model) plus a random error term, say,

$$Y = \beta_0 + \beta_1 x + \epsilon \tag{10-1}$$

where ϵ is the random error term. We will call this model the **simple linear regression model,** because it has only one independent variable or **regressor.** Sometimes a model like this will arise from a theoretical relationship. At other times, we will have no theoretical knowledge of the relationship between x and y, and the choice of the model is based on inspection of a scatter diagram, such as we did with the oxygen purity data. We then think of the regression model as an **empirical model.**

To gain more insight into this model, suppose that we can fix the value of x and observe the value of the random variable Y. Now if x is fixed, the random component ϵ on the right-hand side of the model in equation 10-1 determines the properties of Y. Suppose that the mean and variance of ϵ are 0 and σ^2, respectively. Then

$$E(Y \mid x) = E(\beta_0 + \beta_1 x + \epsilon) = \beta_0 + \beta_1 x + E(\epsilon) = \beta_0 + \beta_1 x$$

Notice that this is the same relationship that we initially wrote down empirically from inspection of the scatter diagram in Fig. 10-1. The variance of Y given x is

$$V(Y \mid x) = V(\beta_0 + \beta_1 x + \epsilon) = V(\beta_0 + \beta_1 x) + V(\epsilon) = 0 + \sigma^2 = \sigma^2$$

Thus, the true regression model $\mu_{Y|x} = \beta_0 + \beta_1 x$ is a line of mean values; that is, the height of the regression line at any value of x is just the expected value of Y for that x. The slope, β_1, can be interpreted as the change in the mean of Y for a unit change in x. Furthermore, the variability of Y at a particular value of x is determined by the error variance σ^2. This implies that there is a distribution of Y-values at each x and that the variance of this distribution is the same at each x.

For example, suppose that the true regression model relating oxygen purity to hydrocarbon level is $\mu_{Y|x} = 75 + 15x$, and suppose that the variance is $\sigma^2 = 2$. Figure 10-2 illustrates this situation. Notice that we have used a normal distribution to describe the random variation in ϵ. Since Y is the sum of a constant $\beta_0 + \beta_1 x$ (the mean) and a normally distributed random variable, Y is a normally distributed random variable. The variance σ^2 determines the variability in the observations Y on oxygen purity. Thus, when σ^2 is small, the observed values of Y will fall close to the line, and when σ^2 is large, the observed values of Y may deviate considerably from the line. Because σ^2 is constant, the variability in Y at any value of x is the same.

The regression model describes the relationship between oxygen purity Y and hydrocarbon level x. Thus, for any value of hydrocarbon level, oxygen purity has a normal distribution with mean $75 + 15x$ and variance 2. For example, if $x = 1.25$, then Y has mean value $\mu_{Y|x} = 75 + 15(1.25) = 93.75$ and variance 2.

In most real-world problems, the values of the intercept and slope (β_0, β_1) and the error variance σ^2 will not be known, and they must be estimated from sample data. **Regression analysis** is a collection of statistical tools for finding estimates of the parameters in the regression model. Then this fitted regression equation or model is typically used in prediction of future observations of Y, or for estimating the mean response at a particular level of x. To illustrate, a chemical engineer might be interested in estimating the mean purity

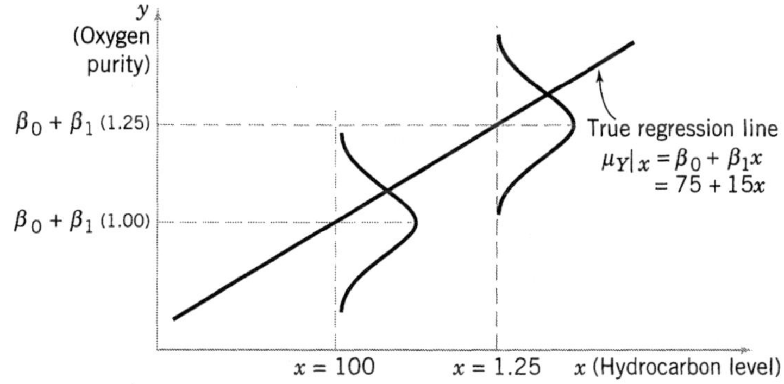

Figure 10-2 The distribution of Y for a given value of x for the oxygen purity–hydrocarbon data.

of oxygen produced when the hydrocarbon level is $x = 1.25\%$. This chapter discusses such procedures and applications for the simple linear regression model. Chapter 11 will discuss multiple linear regression models that involve more than one regressor.

10-2 SIMPLE LINEAR REGRESSION

The case of **simple linear regression** considers a *single regressor* or *predictor* x and a dependent or *response* variable Y. Suppose that the true relationship between Y and x is a straight line and that the observation Y at each level of x is a random variable. As noted previously, the expected value of Y for each value of x is

$$E(Y \mid x) = \beta_0 + \beta_1 x$$

where the intercept β_0 and the slope β_1 are unknown regression coefficients. We assume that each observation, Y, can be described by the model

$$Y = \beta_0 + \beta_1 x + \epsilon \qquad (10\text{-}2)$$

where ϵ is a random error with mean zero and variance σ^2. The random errors corresponding to different observations are also assumed to be uncorrelated random variables.

Suppose that we have n pairs of observations (x_1, y_1), (x_2, y_2), ... (x_n, y_n). Figure 10-3 shows a typical scatter plot of observed data and a candidate for the estimated regression line. The estimates of β_0 and β_1 should result in a line that is (in some sense) a "best fit" to the data. The German scientist Karl Gauss (1777–1855) proposed estimating the parameters β_0 and β_1 in equation 10-2 in order to minimize the sum of the squares of the vertical deviations in Fig. 10-3.

We call this criterion for estimating the regression coefficients the **method of least squares.** Using equation 10-2, we may express the n observations in the sample as

$$y_i = \beta_0 + \beta_1 x_i + \epsilon_i, \qquad i = 1, 2, \ldots, n \qquad (10\text{-}3)$$

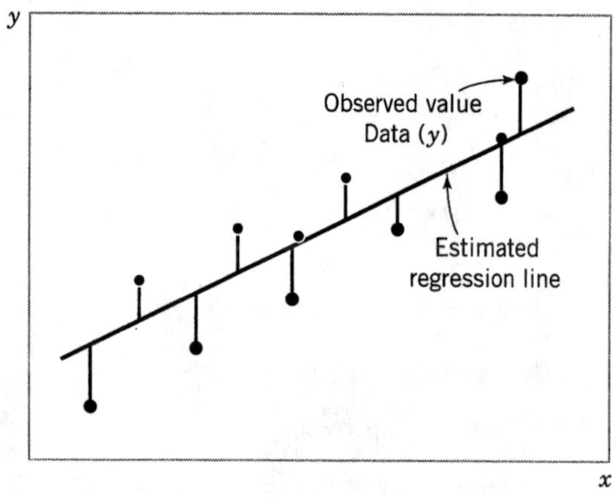

Figure 10-3 Deviations of the data from the estimated regression model.

and the sum of the squares of the deviations of the observations from the true regression line is

$$L = \sum_{i=1}^{n} \epsilon_i^2 = \sum_{i=1}^{n} (y_i - \beta_0 - \beta_1 x_i)^2 \tag{10-4}$$

The least squares estimators of β_0 and β_1, say, $\hat{\beta}_0$ and $\hat{\beta}_1$, must satisfy

$$\begin{aligned} \left.\frac{\partial L}{\partial \beta_0}\right|_{\hat{\beta}_0, \hat{\beta}_1} &= -2 \sum_{i=1}^{n} (y_i - \hat{\beta}_0 - \hat{\beta}_1 x_i) = 0 \\ \left.\frac{\partial L}{\partial \beta_1}\right|_{\hat{\beta}_0, \hat{\beta}_1} &= -2 \sum_{i=1}^{n} (y_i - \hat{\beta}_0 - \hat{\beta}_1 x_i) x_i = 0 \end{aligned} \tag{10-5}$$

Simplifying these two equations yields

$$\begin{aligned} n\hat{\beta}_0 + \hat{\beta}_1 \sum_{i=1}^{n} x_i &= \sum_{i=1}^{n} y_i \\ \hat{\beta}_0 \sum_{i=1}^{n} x_i + \hat{\beta}_1 \sum_{i=1}^{n} x_i^2 &= \sum_{i=1}^{n} y_i x_i \end{aligned} \tag{10-6}$$

Equations 10-6 are called the **least squares normal equations.** The solution to the normal equations results in the least squares estimators $\hat{\beta}_0$ and $\hat{\beta}_1$.

Definition

The **least squares estimates** of the intercept and slope in the simple linear regression model are

$$\hat{\beta}_0 = \bar{y} - \hat{\beta}_1 \bar{x} \tag{10-7}$$

$$\hat{\beta}_1 = \frac{\displaystyle\sum_{i=1}^{n} y_i x_i - \frac{\left(\sum_{i=1}^{n} y_i\right)\left(\sum_{i=1}^{n} x_i\right)}{n}}{\displaystyle\sum_{i=1}^{n} x_i^2 - \frac{\left(\sum_{i=1}^{n} x_i\right)^2}{n}} \tag{10-8}$$

where $\bar{y} = (1/n)\sum_{i=1}^{n} y_i$ and $\bar{x} = (1/n)\sum_{i=1}^{n} x_i$.

The **fitted** or **estimated regression line** is therefore

$$\hat{y} = \hat{\beta}_0 + \hat{\beta}_1 x \tag{10-9}$$

Note that each pair of observations satisfies the relationship

$$y_i = \hat{\beta}_0 + \hat{\beta}_1 x_i + e_i, \qquad i = 1, 2, \ldots, n$$

where $e_i = y_i - \hat{y}_i$ is called the **residual.** The residual describes the error in the fit of the model to the ith observation y_i. Later in this chapter we will use the residuals to provide information about the adequacy of the fitted model.

Notationally, it is occasionally convenient to give special symbols to the numerator and denominator of equation 10-8. Given data $(x_1, y_1), (x_2, y_2), \ldots, (x_n, y_n)$, let

$$S_{xx} = \sum_{i=1}^{n} (x_i - \bar{x})^2 = \sum_{i=1}^{n} x_i^2 - \frac{\left(\sum_{i=1}^{n} x_i\right)^2}{n} \tag{10-10}$$

and

$$S_{xy} = \sum_{i=1}^{n} y_i(x_i - \bar{x})^2 = \sum_{i=1}^{n} x_i y_i - \frac{\left(\sum_{i=1}^{n} x_i\right)\left(\sum_{i=1}^{n} y_i\right)}{n} \tag{10-11}$$

EXAMPLE 10-1

We will fit a simple linear regression model to the oxygen purity data in Table 10-1. The following quantities may be computed:

$$n = 20 \quad \sum_{i=1}^{20} x_i = 23.92 \quad \sum_{i=1}^{20} y_i = 1{,}843.21 \quad \bar{x} = 1.20 \quad \bar{y} = 92.16$$

$$\sum_{i=1}^{20} y_i^2 = 170{,}044.53 \quad \sum_{i=1}^{20} x_i^2 = 29.29 \quad \sum_{i=1}^{20} x_i y_i = 2{,}214.66$$

$$S_{xx} = \sum_{i=1}^{20} x_i^2 - \frac{\left(\sum_{i=1}^{20} x_i\right)^2}{20} = 29.29 - \frac{(23.92)^2}{20} = 0.68$$

and

$$S_{xy} = \sum_{i=1}^{20} x_i y_i - \frac{\left(\sum_{i=1}^{20} x_i\right)\left(\sum_{i=1}^{20} y_i\right)}{20} = 2{,}214.66 - \frac{(23.92)(1{,}843.21)}{20} = 10.18$$

Therefore, the least squares estimates of the slope and intercept are

$$\hat{\beta}_1 = \frac{S_{xy}}{S_{xx}} = \frac{10.18}{0.68} = 14.97$$

and

$$\hat{\beta}_0 = \bar{y} - \hat{\beta}_1\bar{x} = 92.16 - (14.97)1.20 = 74.20$$

The fitted simple linear regression model is

$$\hat{y} = 74.20 + 14.97x$$

This model is plotted in Fig. 10-4, along with the sample data. We have carried two decimal places in the calculation of these regression coefficients.

Computer software programs are widely used in regression modeling. These programs typically carry more decimal places in the calculations. Table 10-2 shows a portion of the output from Minitab for this problem. Notice that the estimates of the slope and intercept are reported to three decimal places and do not exactly match the hand calculations because we carried fewer decimals in the arithmetic than the computer. It would be necessary to carry at least four decimal places in the manual calculations to match the computer output. In subsequent sections we will provide explanations for the other information provided in this computer output.

Using the regression model of Example 10-1 we would predict oxygen purity of $\hat{y} = 89.17\%$ when the hydrocarbon level is $x = 1.00\%$. The purity 89.17% may be

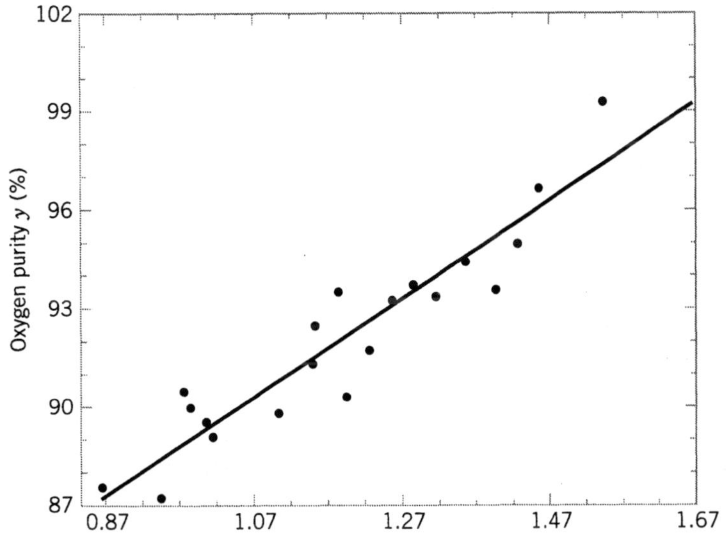

Figure 10-4 Scatter plot of oxygen purity y versus hydrocarbon level x and regression model $\hat{y} = 74.20 + 14.97x$.

Table 10-2 Minitab Output for the Oxygen Purity Data in Example 10-1

Regression Analysis

The regression equation is

$y = 74.3 + 14.9x$

Predictor	Coef	St.Dev	T	P
Constant	74.283	1.593	46.62	0.000
x	14.947	1.317	11.35	0.000

S = 1.087 R-Sq = 87.7% R-Sq(adj) = 87.1%

Analysis of Variance

Source	DF	SS	MS	F	P
Regression	1	152.13	152.13	128.86	0.000
Error	18	21.25	1.18		
Total	19	173.38			

interpreted as an estimate of the true population mean purity when $x = 1.00\%$, or as an estimate of a new observation when $x = 1.00\%$. These estimates are, of course, subject to error; that is, it is unlikely that a future observation on purity would be exactly 89.17% when the hydrocarbon level is 1.00%. In subsequent sections we will see how to use confidence intervals and prediction intervals to describe the error in estimation from a regression model.

EXERCISES FOR SECTION 10-2

10-1. An article in *Concrete Research* ("Near Surface Characteristics of Concrete: Intrinsic Permeability," Vol. 41, 1989), presented data on compressive strength x and intrinsic permeability y of various concrete mixes and cures. Summary quantities are $n = 14$, $\sum y_i = 572$, $\sum y_i^2 = 23{,}530$, $\sum x_i = 43$, $\sum x_i^2 = 157.42$, and $\sum x_i y_i = 1{,}697.80$. Assume that the two variables are related according to the simple linear regression model.

(a) Calculate the least squares estimates of the slope and intercept.

(b) Use the equation of the fitted line to predict what permeability would be observed when the compressive strength is $x = 4.3$.

(c) Give a point estimate of the mean permeability when compressive strength is $x = 3.7$.

(d) Suppose that the observed value of permeability at $x = 3.7$ is $y = 46.1$. Calculate the value of the corresponding residual.

10-2. Regression methods were used to analyze the data from a study investigating the relationship between roadway surface temperature (x) and pavement deflection (y). Summary quantities were $n = 20$, $\Sigma y_i = 12.75$, $\Sigma y_i^2 = 8.86$, $\Sigma x_i = 1478$, $\Sigma x_i^2 = 143{,}215.8$, and $\Sigma x_i y_i = 1083.67$.

(a) Calculate the least squares estimates of the slope and intercept. Graph the regression line.

(b) Use the equation of the fitted line to predict what pavement deflection would be observed when the surface temperature is 85°F.

(c) What is the mean pavement deflection when the surface temperature is 90°F?

(d) What change in mean pavement deflection would be expected for a 1°F change in surface temperature?

10-3. Consider the regression model developed in Exercise 10-2.

(a) Suppose that temperature is measured in °C rather than °F. Write the new regression model that results.

(b) What change in expected pavement deflection is associated with a 1°C change in surface temperature?

10-4. Montgomery and Peck (1992) present data concerning the performance of the 28 National Football League teams in 1976. It is suspected that the number of games won (y) is related to the number of yards gained rushing by an opponent (x). The data are shown in the following table.

(a) Calculate the least squares estimates of the slope and intercept. Graph the regression model.

(b) Find an estimate of the mean number of games won if the opponents can be limited to 1800 yards rushing.

(c) What change in the expected number of games won is associated with a decrease of 100 yards rushing by an opponent?

Teams	Games Won (y)	Yards Rushing by Opponent (x)
Washington	10	2205
Minnesota	11	2096
New England	11	1847
Oakland	13	1903
Pittsburgh	10	1457
Baltimore	11	1848
Los Angeles	10	1564
Dallas	11	1821
Atlanta	4	2577
Buffalo	2	2476
Chicago	7	1984
Cincinnati	10	1917
Cleveland	9	1761
Denver	9	1709
Detroit	6	1901
Green Bay	5	2288
Houston	5	2072
Kansas City	5	2861
Miami	6	2411
New Orleans	4	2289
New York Giants	3	2203
New York Jets	3	2592
Philadelphia	4	2053
St. Louis	10	1979
San Diego	6	2048
San Francisco	8	1786
Seattle	2	2876
Tampa Bay	0	2560

(d) To increase by one the mean number of games won how much decrease in rushing yards must be generated by the defense?

(e) Given that $x = 1917$ yards (Cincinnati), find the fitted value of y and the corresponding residual.

10-5. An article in *Technometrics* by S. C. Narula and J. F. Wellington ("Prediction, Linear Regression, and a Minimum Sum of Relative Errors," Vol. 19, 1977) presents data on the selling price and annual taxes for 24 houses. The data are shown in the following table.

Sale Price/1000	Taxes (Local, School, County)/1000
25.9	4.9176
29.5	5.0208
27.9	4.5429
25.9	4.5573
29.9	5.0597
29.9	3.8910
30.9	5.8980
28.9	5.6039
35.9	5.8282
31.5	5.3003
31.0	6.2712
30.9	5.9592
30.0	5.0500
36.9	8.2464
41.9	6.6969
40.5	7.7841
43.9	9.0384
37.5	5.9894
37.9	7.5422
44.5	8.7951
37.9	6.0831
38.9	8.3607
36.9	8.1400
45.8	9.1416

(a) Assuming that a simple linear regression model is appropriate, obtain the least squares fit relating selling price to taxes paid.

(b) Find the mean selling price given that the taxes paid are $x = 7.50$.

(c) Calculate the fitted value of y corresponding to $x = 5.8980$. Find the corresponding residual.

(d) Calculate the fitted value $\hat{y}_i$ for each value of x_i used to fit the model. Then construct a graph of $\hat{y}_i$ versus the corresponding observed value y_i and comment on what this plot would look like if the relationship between y and x was a deterministic (no random error) straight line. Does the plot actually obtained indicate that taxes paid is an effective regressor variable in predicting selling price?

10-6. The number of pounds of steam used per month by a chemical plant is thought to be related to the average ambient temperature (in °F) for that month. The past year's usage and temperature are shown in the following table:

Month	Temp.	Usage/1000
Jan.	21	185.79
Feb.	24	214.47
Mar.	32	288.03
Apr.	47	424.84
May	50	454.58
June	59	539.03
July	68	621.55
Aug.	74	675.06
Sept.	62	562.03
Oct.	50	452.93
Nov.	41	369.95
Dec.	30	273.98

(a) Assuming that a simple linear regression model is appropriate, fit the regression model relating steam usage (y) to the average temperature (x).

(b) What is the estimate of expected steam usage when the average temperature is 55°F?

(c) What change in mean steam usage is expected when the monthly average temperature changes by 1°F?

(d) Suppose the monthly average temperature is 47°F. Calculate the fitted value of y and the corresponding residual.

10-7. The data shown in the following table are highway gasoline mileage performance and engine displacement for a sample of 20 cars.

Make	Model	MPG (highway)	Engine Displacement (in^3)
Acura	Legend	30	97
BMW	735i	19	209
Buick	Regal	29	173
Chevrolet	Cavalier	32	121
Chevrolet	Celebrity	30	151
Chrysler	Conquest	24	156
Dodge	Aries	30	135
Dodge	Dynasty	28	181
Ford	Escort	31	114
Ford	Mustang	25	302
Ford	Taurus	27	153
Ford	Tempo	33	90
Honda	Accord	30	119
Mazda	RX-7	23	80
Mercedes	260E	24	159
Mercury	Tracer	29	97
Nissan	Maxima	26	181
Oldsmobile	Cutlass	29	173
Plymouth	Laser	37	122
Pontiac	Grand Prix	29	173

(a) Fit a simple linear model relating highway MPG (y) to engine displacement (x) using least squares.

(b) Find an estimate of the mean highway gasoline mileage performance for a car with 150 in^3 engine displacement.

(c) Obtain the fitted value of y and the corresponding residual for a car, the Ford Escort, with engine displacement of 114 in^3.

10-8. An article in the *Tappi Journal* (March, 1986) presented data on green liquor Na_2S concentration and paper machine production. The data (read from a graph) are shown in the following table.

Observation Number	Green Liquor Na_2S Concentration (g/l)	Production (tons/day)
1	40	825
2	42	830
3	49	890
4	46	895
5	44	890
6	48	910
7	46	915
8	43	960
9	53	990
10	52	1010
11	54	1012
12	57	1030
13	58	1050

(a) Fit a simple linear regression model with y-green liquor Na_2S concentration and x-production. Draw a scatter diagram of the data and the resulting least squares fitted model.

(b) Find the fitted value of y corresponding to $x = 910$ and the associated residual.

(c) Find the mean green liquor Na_2S concentration when the production rate is 950 tons per day.

10-9. An article in the *Journal of Sound and Vibration* (Vol. 151, 1991, pp. 383–394) described a study investigating the relationship between noise exposure and hypertension. The following data are representative of those reported in the article.

(a) Draw a scatter diagram of y (blood pressure rise) versus x (sound pressure level). Does a simple linear regression model seem reasonable in this situation?

(b) Fit the simple linear regression model using least squares.

Evoked Blood Pressure Rise (mmHg)	Sound Pressure Level (dB)
1	60
0	63
1	65
2	70
5	70
1	70
4	80
6	80
2	80
3	80
5	85
4	89
6	90
8	90
4	90
5	90
7	94
9	100
7	100
6	100

(c) Find the predicted mean rise in blood pressure level associated with a sound pressure level of 85 dB.

10-10. An article in *Wear* (Vol. 152, 1992, pp. 171–181) presents data on the fretting wear of mild steel and oil viscosity. Representative data follow, with x = oil viscosity and y = wear volume (10^{-4} mm^3).

y	240	181	193	155	172
x	1.6	9.4	15.5	20.0	22.0

y	110	113	75	94
x	35.5	43.0	40.5	33.0

(a) Construct a scatter plot of the data. Does a simple linear regression model appear to be plausible?

(b) Fit the simple linear regression model using least squares.

(c) Predict fretting wear when viscosity $x = 30$.

(d) Obtain the fitted value of y when $x = 22.0$ and calculate the corresponding residual.

10-11. An article in the *Journal of Environmental Engineering* (Vol. 115, No. 3, 1989, pp. 608–619) reported the results of a study on the occurrence of sodium and chloride in surface streams in central Rhode Island. The data below are chloride concentration y (in mg/l) and roadway area in the watershed x (in %).

y	4.4	6.6	9.7	10.6	10.8
x	0.19	0.15	0.57	0.70	0.67

y	10.9	11.8	12.1	14.3	14.7
x	0.63	0.47	0.70	0.60	0.78

y	15.0	17.3	19.2	23.1	27.4
x	0.81	0.78	0.69	1.30	1.05

y	27.7	31.8	39.5
x	1.06	1.74	1.62

(a) Draw a scatter diagram of the data. Does a simple linear regression model seem appropriate here?

(b) Fit the simple linear regression model using the method of least squares.

(c) Estimate the mean chloride concentration for a watershed that has 1% roadway area.

(d) Find the fitted value corresponding to $x = 0.47$ and the associated residual.

10-12. A rocket motor is manufactured by bonding together two types of propellants, an igniter and a sustainer. The shear strength of the bond y is thought to be a linear function of the age of the propellant x when the motor is cast. Twenty observations are shown in the following table.

Observation Number	Strength y (psi)	Age x (weeks)
1	2158.70	15.50
2	1678.15	23.75
3	2316.00	8.00
4	2061.30	17.00
5	2207.50	5.00
6	1708.30	19.00
7	1784.70	24.00
8	2575.00	2.50
9	2357.90	7.50
10	2277.70	11.00
11	2165.20	13.00
12	2399.55	3.75
13	1779.80	25.00
14	2336.75	9.75
15	1765.30	22.00
16	2053.50	18.00
17	2414.40	6.00
18	2200.50	12.50
19	2654.20	2.00
20	1753.70	21.50

(a) Draw a scatter diagram of the data. Does the straight-line regression model seem to be plausible?

(b) Find the least squares estimates of the slope and intercept in the simple linear regression model.

(c) Estimate the mean shear strength of a motor made from propellant that is 20 weeks old.

(d) Obtain the fitted values $\hat{y}_i$ that correspond to each observed value y_i. Plot $\hat{y}_i$ versus y_i, and comment on what this plot would look like if the linear relationship between shear strength and age were perfectly deterministic (no error). Does this plot indicate that age is a reasonable choice of regressor variable in this model?

10-13. Show that in a simple linear regression model the point $(\bar{x}, \bar{y})$ lies exactly on the least squares regression line.

10-14. Consider the simple linear regression model $Y = \beta_0 + \beta_1 x + \epsilon$. Suppose that the analyst wants to use $z = x - \bar{x}$ as the regressor variable.

(a) Using the data in Exercise 10-12, construct one scatter plot of the (x_i, y_i) points and then another of the $(z_i = x_i - \bar{x}, y_i)$ points. Use the two plots to intuitively explain how the two models, $Y = \beta_0 + \beta_1 x + \epsilon$ and $Y = \beta_0^* + \beta_1^* z + \epsilon$, are related.

(b) Find the least squares estimates of β_0^* and β_1^* in the model $Y = \beta_0^* + \beta_1^* z + \epsilon$. How do they relate to the least squares estimates $\hat{\beta}_0$ and $\hat{\beta}_1$?

10-15. Suppose we wish to fit the model $Y_i^* = \beta_0^* + \beta_1^*(x_i - \bar{x}) + \epsilon_i$, where $Y_i^* = Y_i - \bar{Y}$ $(i = 1, 2, \ldots, n)$. Find the least squares estimates of β_0^* and β_1^*. How do they relate to $\hat{\beta}_0$ and $\hat{\beta}_1$?

10-16. Suppose we wish to fit a regression model for which the true regression line passes through the point (0, 0). The appropriate model is $Y = \beta x + \epsilon$. Assume that we have n pairs of data $(x_1, y_1), (x_2, y_2), \ldots, (x_n, y_n)$. Find the least squares estimate of β.

10-17. Using the results of Exercise 10-16, fit the model $Y = \beta x + \epsilon$ to the chloride concentration-roadway area data in Exercise 10-11. Plot the fitted model on a scatter diagram of the data and comment on the appropriateness of the model.

10-3 PROPERTIES OF THE LEAST SQUARES ESTIMATORS AND ESTIMATION OF σ^2

The statistical properties of the least squares estimators $\hat{\beta}_0$ and $\hat{\beta}_1$ may be easily described. Recall that we have assumed that the error term ϵ in the model $Y = \beta_0 + \beta_1 x + \epsilon$ is a random variable with mean zero and variance σ^2. Since the values of x are fixed, Y is a random variable with mean $\mu_{Y|x} = \beta_0 + \beta_1 x$ and variance σ^2. Therefore, the values of $\hat{\beta}_0$ and $\hat{\beta}_1$ depend on the observed y's; thus, the least squares estimators of the regression coefficients may be viewed as random variables. We will investigate the bias and variance properties of the least squares estimators $\hat{\beta}_0$ and $\hat{\beta}_1$.

Consider first $\hat{\beta}_1$. Because $\hat{\beta}_1$ is a linear combination of the observations Y_i, we can use the results from Chapter 6 to show that expected value of $\hat{\beta}_1$ is

$$E(\hat{\beta}_1) = \beta_1 \tag{10-12}$$

Thus, $\hat{\beta}_1$ is an **unbiased estimator** of the true slope β_1.

Now consider the variance of $\hat{\beta}_1$. Since we have assumed that $V(\epsilon_i) = \sigma^2$, it follows that $V(Y_i) = \sigma^2$, and it can be shown that

$$V(\hat{\beta}_1) = \frac{\sigma^2}{S_{xx}} \tag{10-13}$$

By using a similar approach, we can show that

$$E(\hat{\beta}_0) = \beta_0 \quad \text{and} \quad V(\hat{\beta}_0) = \sigma^2\left[\frac{1}{n} + \frac{\overline{x}^2}{S_{xx}}\right] \tag{10-14}$$

Thus, $\hat{\beta}_0$ is an unbiased estimator of the intercept β_0. The covariance of the random variables $\hat{\beta}_0$ and $\hat{\beta}_1$ is not zero. It can be shown (see Exercise 10-69) that $\text{cov}(\hat{\beta}_0, \hat{\beta}_1) = -\sigma^2\overline{x}/S_{xx}$.

To draw inferences concerning the regression coefficients β_0 and β_1, it is necessary to estimate the variance σ^2 appearing in the expressions for $V(\hat{\beta}_0)$ and $V(\hat{\beta}_1)$. The parameter σ^2, the variance of the error term ϵ in the regression model, reflects random variation around the true regression line.

The residuals, $e_i = y_i - \hat{y}_i$, are used in computing the estimate of σ^2. The sum of the squares of the residuals, or the **error sum of squares** is

$$SS_E = \sum_{i=1}^{n} e_i^2 = \sum_{i=1}^{n} (y_i - \hat{y}_i)^2 \tag{10-15}$$

It can be shown that the expected value of the error sum of squares SS_E is

$$E(SS_E) = (n - 2)\sigma^2 \tag{10-16}$$

Therefore,

$$\hat{\sigma}^2 = \frac{SS_E}{n-2} \tag{10-17}$$

is an unbiased estimator of σ^2.

A more convenient computing formula for SS_E may be found by substituting the fitted model $\hat{y}_i = \hat{\beta}_0 + \hat{\beta}_1 x_i$ into equation 10-15 and simplifying. The result is

$$SS_E = \sum_{i=1}^{n} y_i^2 - n\bar{y}^2 - \hat{\beta}_1 S_{xy}$$

and if we let $SS_T \equiv \sum_{i=1}^{n} y_i^2 - n\bar{y}^2 = \sum_{i=1}^{n} (y_i - \bar{y})^2$, then we may write SS_E as

$$SS_E = SS_T - \hat{\beta}_1 S_{xy} \tag{10-18}$$

EXAMPLE 10-2

We will find the estimate of the variance σ^2 using the data from Example 10-1. Now $S_{xy} = 10.18$, $\hat{\beta}_1 = 14.97$, and

$$SS_T = \sum_{i=1}^{20} y_i^2 - \frac{\left(\sum_{i=1}^{20} y_i\right)^2}{20} = 170{,}044.53 - \frac{(1{,}843.21)^2}{20}$$
$$= 173.37$$

Therefore, using equation 10-17

$$\hat{\sigma}^2 = \frac{SS_E}{n-2} = \frac{SS_T - \hat{\beta}_1 S_{xy}}{n-2}$$
$$= \frac{173.37 - (14.97)(10.18)}{20-2} = 1.17$$

The estimate of σ^2 ($\hat{\sigma}^2 = 1.17$) could be used in equation 10-14 to provide an estimate of the variance of the slope and in equation 10-15 to estimate the variance of the intercept. We call the square roots of the resulting variance estimators the **estimated standard errors** of the slope and intercept, respectively.

Definition

In simple linear regression the **estimated standard error of the slope** is

$$se(\hat{\beta}_1) = \sqrt{\frac{\hat{\sigma}^2}{S_{xx}}}$$

and the **estimated standard error of the intercept** is

$$se(\hat{\beta}_0) = \sqrt{\hat{\sigma}^2\left[\frac{1}{n} + \frac{\bar{x}^2}{S_{xx}}\right]}$$

where $\hat{\sigma}^2$ is computed from equation 10-17.

The Minitab computer output in Table 10-2 reports the standard errors of the slope and intercept, respectively. Notice that the estimate of σ^2 in that table is listed under the "mean square residual" entry in the analysis of variance portion of the display. (This information will be discussed in Section 10-5.) Moreover, because the computer carries more decimal places in the calculations than we did manually, Minitab calculates the mean square error as 1.18, which differs slightly from the value 1.17 obtained above.

10-4 COMMON ABUSES OF REGRESSION

Regression is widely used and frequently misused; several common abuses of regression are briefly mentioned here. Care should be taken in selecting variables with which to construct regression equations and in determining the form of the model. It is possible to develop statistical relationships among variables that are completely unrelated in a practical sense. For example, we might attempt to relate the shear strength of spot welds with the number of boxes of computer paper used by the information systems group. A straight line may even appear to provide a good fit to the data, but the relationship is an unreasonable one on which to rely. A strong observed association between variables does not necessarily imply that a causal relationship exists between those variables. Designed experiments are the only way to determine causal relationships.

Regression relationships are valid only for values of the regressor variable within the range of the original data. The linear relationship that we have tentatively assumed may be valid over the original range of x, but it may be unlikely to remain so as we extrapolate—that is, if we use values of x beyond that range. In other words, as we move beyond the range of values of x for which data were collected, we become less certain about the validity of the assumed model. Regression models are not necessarily valid for extrapolation purposes.

10-5 HYPOTHESIS TESTS IN SIMPLE LINEAR REGRESSION

An important part of assessing the adequacy of a linear regression model is testing statistical hypotheses about the model parameters and constructing certain confidence intervals. Hypothesis testing in simple linear regression is discussed in this section, and Section 10-6 presents methods for constructing confidence intervals. To test hypotheses about the slope and intercept of the regression model, we must make the additional assumption that the error component in the model, ϵ, is normally distributed. Thus, the complete assumptions are that the errors are normally and independently distributed with mean zero and variance σ^2, abbreviated NID(0, σ^2).

10-5.1 Use of t-Tests

Suppose we wish to test hypotheses that the slope equals a constant, say, $\beta_{1,0}$. The appropriate hypotheses are

$$H_0: \beta_1 = \beta_{1,0}$$
$$H_1: \beta_1 \neq \beta_{1,0} \tag{10-19}$$

where we have assumed a two-sided alternative. Since the errors ϵ_i are NID(0, σ^2), it follows directly that the observations Y_i are NID($\beta_0 + \beta_1 x_i$, σ^2). From equation 10-8 we observe that $\hat{\beta}_1$ is a linear combination of the observations Y_i. Thus, $\hat{\beta}_1$ is a linear combination of independent normal random variables, and consequently, $\hat{\beta}_1$ is N(β_1, σ^2/S_{xx}), using the bias and variance properties of the slope discussed in Section 10-3. In addition, $(n-2)\hat{\sigma}^2/\sigma^2$ has a chi-square distribution with $n-2$ degrees of freedom, and $\hat{\beta}_1$ is independent of $\hat{\sigma}^2$. As a result of those properties, the statistic

$$T_0 = \frac{\hat{\beta}_1 - \beta_{1,0}}{\sqrt{\hat{\sigma}^2/S_{xx}}} \tag{10-20}$$

follows the t distribution with $n-2$ degrees of freedom under H_0: $\beta_1 = \beta_{1,0}$. We would reject H_0: $\beta_1 = \beta_{1,0}$ if

$$|t_0| > t_{\alpha/2,n-2} \tag{10-21}$$

where t_0 is computed from equation 10-20. The denominator of equation 10-20 is the standard error of the slope, so we could write the test statistic as

$$T_0 = \frac{\hat{\beta}_1 - \beta_{1,0}}{se(\hat{\beta}_1)}$$

A similar procedure can be used to test hypotheses about the intercept. To test

$$H_0\colon \beta_0 = \beta_{0,0}$$
$$H_1\colon \beta_0 \neq \beta_{0,0} \qquad (10\text{-}22)$$

we would use the statistic

$$T_0 = \frac{\hat{\beta}_0 - \beta_{0,0}}{\sqrt{\hat{\sigma}^2\left[\frac{1}{n} + \frac{\bar{x}^2}{S_{xx}}\right]}} = \frac{\hat{\beta}_0 - \beta_{0,0}}{se(\hat{\beta}_0)} \qquad (10\text{-}23)$$

and reject the null hypothesis if the computed value of this test statistic, t_0, is such that $|t_0| > t_{\alpha/2, n-2}$. Note that the denominator of the test statistic in equation 10-23 is just the standard error of the intercept.

A very important special case of the hypotheses of equation 10-19 is

$$H_0\colon \beta_1 = 0$$
$$H_1\colon \beta_1 \neq 0 \qquad (10\text{-}24)$$

These hypotheses relate to the **significance of regression.** Failure to reject $H_0\colon \beta_1 = 0$ is equivalent to concluding that there is no linear relationship between x and Y. This situation is illustrated in Fig. 10-5. Note that this may imply either that x is of little value in explaining the variation in Y and that the best estimator of Y for any x is $\hat{Y} = \bar{Y}$ (Fig. 10-5*a*) or that the true relationship between x and Y is not linear (Fig. 10-5*b*).

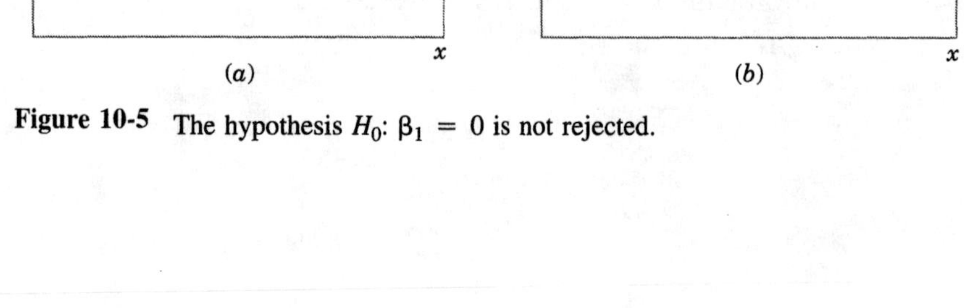

Figure 10-5 The hypothesis $H_0\colon \beta_1 = 0$ is not rejected.

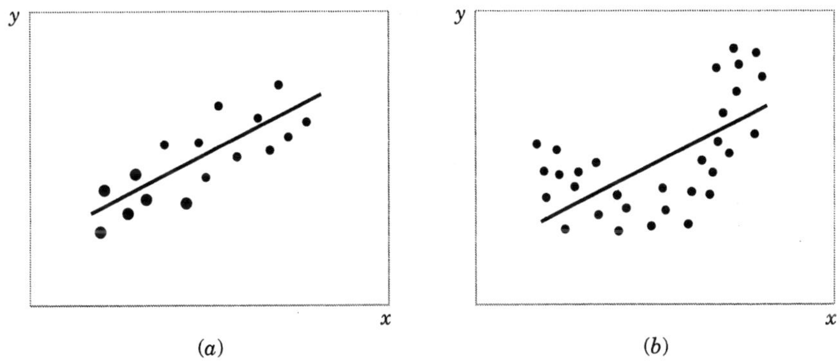

Figure 10-6 The hypothesis H_0: $\beta_1 = 0$ is rejected.

Alternatively, if H_0: $\beta_1 = 0$ is rejected, this implies that x is of value in explaining the variability in Y (see Fig. 10-6). Rejecting H_0: $\beta_1 = 0$ could mean either that the straight-line model is adequate (Fig. 10-6*a*) or that, although there is a linear effect of x, better results could be obtained with the addition of higher order polynomial terms in x (Fig. 10-6*b*).

EXAMPLE 10-3

We will test for significance of regression using the model for the oxygen purity data from Example 10-1. The hypotheses are

$$H_0: \beta_1 = 0$$
$$H_1: \beta_1 \neq 0$$

and we will use $\alpha = 0.01$. From Examples 10-1 and 10-2 we have

$$\hat{\beta}_1 = 14.97 \quad n = 20, \quad S_{xx} = 0.68, \quad \hat{\sigma}^2 = 1.17$$

so the t-statistic in equation 10-20 becomes

$$t_0 = \frac{\hat{\beta}_1}{\sqrt{\hat{\sigma}^2/S_{xx}}} = \frac{\hat{\beta}_1}{se(\hat{\beta}_1)} = \frac{14.97}{\sqrt{1.17/0.68}} = 11.41$$

Since the reference value of t is $t_{0.005,18} = 2.88$, the value of the test statistic is very far into the critical region, implying that H_0: $\beta_1 = 0$ should be rejected. The P-value for this test is $P \simeq 1.13 \times 10^{-9}$. This was obtained manually with a calculator.

As noted earlier, Table 10-2 presents the Minitab output for this problem. Notice that the t-statistic value for the slope is computed as 11.35 and that the reported P-value is $P = 0.00000$. Minitab also reports the t-statistic for testing the hypothesis H_0: $\beta_0 = 0$. This statistic is computed from equation 10-23, with $\beta_{0,0} = 0$, as $t_0 = 46.62$. Clearly, then, the hypothesis that the intercept is zero is rejected.

10-5.2 Analysis of Variance Approach to Test Significance of Regression

A method called the **analysis of variance** can be used to test for significance of regression. The procedure partitions the total variability in the response variable into meaningful components as the basis for the test. The *analysis of variance identity* is as follows:

$$\sum_{i=1}^{n} (y_i - \bar{y})^2 = \sum_{i=1}^{n} (\hat{y}_i - \bar{y})^2 + \sum_{i=1}^{n} (y_i - \hat{y}_i)^2 \tag{10-25}$$

The two components of the right-hand-side of equation 10-25 measure, respectively, the amount of variability in y_i accounted for by the regression line and the residual variation left unexplained by the regression line. We usually call $SS_E = \sum_{i=1}^{n} (y_i - \hat{y}_i)^2$ the **error sum of squares** and $SS_R = \sum_{i=1}^{n} (\hat{y}_i - \bar{y})^2$ the **regression sum of squares.** Symbolically, equation 10-25 may be written as

$$SS_T = SS_R + SS_E \tag{10-26}$$

where $SS_T = \sum_{i=1}^{n} (y_i - \bar{y})^2$ is the **total corrected sum of squares** of y. In Section 10-3 we noted that $SS_E = SS_T - \hat{\beta}_1 S_{xy}$ (see equation 10-18), so since $SS_T = \hat{\beta}_1 S_{xy} + SS_E$ we note that the regression sum of squares in equation 10-26 is

$$SS_R = \hat{\beta}_1 S_{xy} \tag{10-27}$$

The total sum of squares SS_T has $n - 1$ degrees of freedom, and SS_R and SS_E have 1 and $n - 2$ degrees of freedom, respectively.

We may show that $E[SS_E/(n - 2)] = \sigma^2$, $E(SS_R) = \sigma^2 + \beta_1^2 S_{xx}$ and that SS_E/σ^2 and SS_R/σ^2 are independent chi-square random variables with $n - 2$ and 1 degrees of freedom, respectively. Thus, if the null hypothesis H_0: $\beta_1 = 0$ is true, the statistic

$$F_0 = \frac{SS_R/1}{SS_E/(n-2)} = \frac{MS_R}{MS_E} \tag{10-28}$$

follows the $F_{1,n-2}$ distribution, and we would reject H_0 if $f_0 > f_{\alpha,1,n-2}$. The quantities $MS_R = SS_R/1$ and $MS_E = SS_E/(n-2)$ are called **mean squares.** In general, a mean square is always computed by dividing a sum of squares by its number of degrees of freedom. The test procedure is usually arranged in an **analysis of variance table,** such as Table 10-3.

EXAMPLE 10-4

We will use the analysis of variance approach to test for significance of regression using the oxygen purity data model from Example 10-1. Recall that $SS_T = 173.37$, $\hat{\beta}_1 = 14.97$, $S_{xy} = 10.18$, and $n = 20$. The regression sum of squares is

$$SS_R = \hat{\beta}_1 S_{xy} = (14.97)10.18 = 152.39$$

and the error sum of squares is

$$\begin{aligned} SS_E &= SS_T - SS_R \\ &= 173.37 - 152.39 \\ &= 20.98 \end{aligned}$$

The analysis of variance for testing H_0: $\beta_1 = 0$ is summarized in Table 10-4. The test statistic is $f_0 = MS_R/MS_E = 152.39/1.17 = 130.25$, for which we find that the P-value is $P \simeq 1.13 \times 10^{-9}$, so we conclude that β_1 is not zero.

The Minitab output in Table 10-2 also presents the analysis of variance for this problem. There are frequently minor differences in terminology among computer packages. For example, sometimes the regression sum of squares is called the "model" sum of squares, and the error sum of squares is called the "residual" sum of squares. The numerical results obtained from the computer are very similar to those obtained manually above.

Table 10-3 Analysis of Variance for Testing Significance of Regression

Source of Variation	Sum of Squares	Degrees of Freedom	Mean Square	F_0
Regression	$SS_R = \hat{\beta}_1 S_{xy}$	1	MS_R	MS_R/MS_E
Error	$SS_E = SS_T - \hat{\beta}_1 S_{xy}$	$n - 2$	MS_E	
Total	SS_T	$n - 1$		

Note that $MS_E = \hat{\sigma}^2$.

Table 10-4 Testing for Significance of Regression, Example 10-4

Source of Variation	Sum of Squares	Degrees of Freedom	Mean Square	f_0	P-value
Regression	152.39	1	152.39	130.25	1.13×10^{-9}
Error	20.98	18	1.17		
Total	173.37	19			

Note that the analysis of variance procedure for testing for significance of regression is equivalent to the t-test in Section 10-5.1. That is, either procedure will lead to the same conclusions. This is easy to demonstrate by starting with the t-test statistic in equation 10-20 with $\beta_{1,0} = 0$, say

$$T_0 = \frac{\hat{\beta}_1}{\sqrt{\hat{\sigma}^2/S_{xx}}} \tag{10-29}$$

Squaring both sides of equation 10-29, and using the fact that $\hat{\sigma}^2 = MS_E$ results in

$$T_0^2 = \frac{\hat{\beta}_1^2 S_{xx}}{MS_E} = \frac{\hat{\beta}_1 S_{xY}}{MS_E} = \frac{MS_R}{MS_E} \tag{10-30}$$

Note that T_0^2 in equation 10-30 is identical to F_0 in equation 10-28. It is true, in general, that the square of a t-random variable with ν degrees of freedom is an F-random variable, with one and ν degrees of freedom in the numerator and denominator, respectively. Thus, the test using T_0 is equivalent to the test based on F_0. Note, however, that the t-test is somewhat more flexible in that it would allow testing against a one-sided alternative hypothesis, while the F-test is restricted to a two-sided alternative.

EXERCISES FOR SECTION 10-5

10-18. Consider the data from Exercise 10-1 on x = compressive strength and y = intrinsic permeability of concrete.

(a) Test for significance of regression using $\alpha = 0.05$. Find the P-value for this test. Can you conclude that the model specifies a useful linear relationship between these two variables?

(b) Estimate σ^2 and the standard deviation of $\hat{\beta}_1$.

(c) What is the standard error of the intercept in this model?

10-19. Consider the data from Exercise 10-2 on x = roadway surface temperature and y = pavement deflection.

(a) Test for significance of regression using $\alpha = 0.05$. Find the P-value for this test. What conclusions can you draw?

(b) Estimate σ^2.

(c) Estimate the standard errors of the slope and intercept.

10-20. Consider the National Football League data in Exercise 10-4.

(a) Test for significance of regression

using $\alpha = 0.01$. Find the P-value for this test. What conclusions can you draw?

(b) Estimate σ^2.

(c) Estimate the standard errors of the slope and intercept.

(d) Test (using $\alpha = 0.01$) H_0: $\beta_1 = -0.01$ versus H_1: $\beta_1 \neq -0.01$. Would you agree with the statement that this is a test of the claim that if you can decrease the opponent's rushing yardage by 100 yards the team will win one more game?

10-21. Consider the data from Exercise 10-5 on y = sales price and x = taxes paid.

(a) Test H_0: $\beta_1 = 0$ using the t-test; use $\alpha = 0.05$.

(b) Test H_0: $\beta_1 = 0$ using the analysis of variance with $\alpha = 0.05$. Discuss the relationship of this test to the test from part (a) above.

(c) Estimate σ^2, and estimate the standard errors of the slope and intercept.

(d) Test the hypothesis that $\beta_0 = 0$.

10-22. Consider the data from Exercise 10-6 on y = steam usage and x = average temperature.

(a) Test for significance of regression using $\alpha = 0.01$. What is the P-value for this test? State the conclusions that result from this test.

(b) Estimate σ^2.

(c) Estimate the standard errors of the slope and intercept.

(d) Test the hypothesis H_0: $\beta_1 = 10$ versus H_1: $\beta_1 \neq 10$ using $\alpha = 0.01$. Find the P-value for this test.

(e) Test H_0: $\beta_0 = 0$ versus H_1: $\beta_0 \neq 0$ using $\alpha = 0.01$. Find the P-value for this test and draw conclusions.

10-23. Exercise 10-7 gave 20 observations on y = highway gasoline mileage and x = engine displacement.

(a) Test for significance of regression using $\alpha = 0.01$. Find the P-value for this test. What conclusions can you reach?

(b) Estimate σ and the standard errors of the slope and intercept.

(c) Test H_0: $\beta_1 = -0.05$ versus H_1: $\beta_1 < -0.05$ using $\alpha = 0.01$ and draw conclusions. What is the P-value for this test?

(d) Test the hypothesis H_0: $\beta_0 = 0$ versus H_1: $\beta_0 \neq 0$ using $\alpha = 0.01$. What is the P-value for this test?

10-24. Exercise 10-8 gave 13 observations on y = green liquor Na_2S concentration and x = production in a paper mill.

(a) Test for significance of regression using $\alpha = 0.05$. Find the P-value for this test.

(b) Estimate σ and the standard errors of the slope and intercept.

(c) Test H_0: $\beta_0 = 0$ versus H_1: $\beta_0 \neq 0$ using $\alpha = 0.05$. What is the P-value for this test?

10-25. Exercise 10-9 presented data on y = blood pressure rise and x = sound pressure level.

(a) Test for significance of regression using $\alpha = 0.05$. What is the P-value for this test?

(b) Estimate σ^2.

(c) Estimate the standard errors of the slope and intercept.

(d) Test H_0: $\beta_0 = 0$ versus H_1: $\beta_0 \neq 0$ using $\alpha = 0.05$. Find the P-value for this test.

10-26. Exercise 10-11 presented data on y = chloride concentration in surface streams and x = roadway area.

(a) Test the hypothesis H_0: $\beta_1 = 0$ versus H_1: $\beta_1 \neq 0$ using the analysis of variance procedure with $\alpha = 0.01$.

(b) Find the P-value for the test in part (a).

(c) Estimate σ^2 and the standard errors of the $\hat{\beta}_1$ and $\hat{\beta}_0$.

(d) Test H_0: $\beta_0 = 0$ versus H_1: $\beta_0 \neq 0$ using $\alpha = 0.01$. What conclusions can you draw? Does it seem that the model might be a better fit to the data if the intercept were removed?

10-27. Refer to Exercise 10-12, which gives 20 observations on y = shear strength of a propellant and x = propellant age.

(a) Test for significance of regression with $\alpha = 0.01$. Find the P-value for this test.

(b) Estimate σ^2 and the standard errors of $\hat{\beta}_0$ and $\hat{\beta}_1$.

(c) Test H_0: $\beta_1 = -30$ versus H_1: $\beta_1 \neq -30$ using $\alpha = 0.01$. What is the P-value for this test?

(d) Test H_0: $\beta_0 = 0$ versus H_1: $\beta_0 \neq 0$ using $\alpha = 0.01$. What is the P-value for this test?

(e) Test H_0: $\beta_0 = 2500$ versus H_1: $\beta_0 > 2500$ using $\alpha = 0.01$. What is the P-value for this test?

10-28. Suppose that each value of x_i is multiplied by a positive constant a, and each value of y_i is multiplied by another positive constant b. Show that the t-statistic for testing H_0: $\beta_1 = 0$ versus H_1: $\beta_1 \neq 0$ is unchanged in value.

10-29. Consider the no-intercept model $Y = \beta x + \epsilon$ with the ϵ's NID$(0, \sigma^2)$. The estimate of σ^2 is $s^2 = \sum_{i=1}^{n}(y_i - \hat{\beta}x_i)^2/(n-1)$ and $V(\hat{\beta}) = \sigma^2/\sum_{i=1}^{n} x_i^2$.

(a) Devise a test statistic for H_0: $\beta = 0$ versus H_1: $\beta \neq 0$.

(b) Apply the test in (a) to the model from Exercise 10-17.

10-30. The type II error probability for the t-test for H_0: $\beta_1 = \beta_{1,0}$ can be computed in a similar manner to the t-tests of Chapter 9. If the true value of β_1 is β_1', the value

$$d = \frac{|\beta_{1,0} - \beta_1'|}{\sigma\sqrt{\dfrac{n-1}{S_{xx}}}}$$

is calculated and used as the horizontal scale factor on the operating characteristic curves for the t-test, (Appendix Charts VI*e* through VI*h*) and the type II error probability is read from the vertical scale using the curve for $n - 2$ degrees of freedom. Apply this procedure to the football data of Exercise 10-4, using $\sigma = 2.4$ and $\beta_1' = -0.005$, where the hypotheses are H_0: $\beta_1 = -0.01$ versus H_1: $\beta_1 \neq -0.01$.

10-6 CONFIDENCE INTERVALS

10-6.1 Confidence Intervals on the Slope and Intercept

In addition to point estimates of the slope and intercept, it is possible to obtain confidence interval estimates of these parameters. The width of these confidence intervals is a measure of the overall quality of the regression line. If the error terms, ϵ_i, in the regression model are normally and independently distributed, then

$$(\hat{\beta}_1 - \beta_1)/\sqrt{\hat{\sigma}^2/S_{xx}} \quad \text{and} \quad (\hat{\beta}_0 - \beta_0)\Big/\sqrt{\hat{\sigma}^2\left[\frac{1}{n} + \frac{\bar{x}^2}{S_{xx}}\right]}$$

are both distributed as t-random variables with $n - 2$ degrees of freedom. This leads to the following definition of $100(1 - \alpha)$ percent confidence intervals on the slope and intercept.

Definition

Under the assumption that the observations are normally and independently distributed, a $100(1 - \alpha)$ percent **confidence interval on the slope** β_1 **in simple linear regression** is

$$\hat{\beta}_1 - t_{\alpha/2,n-2}\sqrt{\frac{\hat{\sigma}^2}{S_{xx}}} \le \beta_1 \le \hat{\beta}_1 + t_{\alpha/2,n-2}\sqrt{\frac{\hat{\sigma}^2}{S_{xx}}} \tag{10-31}$$

Similarly, a $100(1 - \alpha)$ percent **confidence interval on the intercept** β_0 is

$$\hat{\beta}_0 - t_{\alpha/2,n-2}\sqrt{\hat{\sigma}^2\left[\frac{1}{n} + \frac{\bar{x}^2}{S_{xx}}\right]} \le \beta_0 \le \hat{\beta}_0 + t_{\alpha/2,n-2}\sqrt{\hat{\sigma}^2\left[\frac{1}{n} + \frac{\bar{x}^2}{S_{xx}}\right]} \tag{10-32}$$

EXAMPLE 10-5

We will find a 95% confidence interval on the slope of the regression line using the data in Example 10-1. Recall that $\hat{\beta}_1 = 14.97$, $S_{xx} = 0.68$, and $\hat{\sigma}^2 = 1.17$ (see Table 10-2). Then, from equation 10-31 we find

$$\hat{\beta}_1 - t_{0.025,18}\sqrt{\frac{\hat{\sigma}^2}{S_{xx}}} \le \beta_1 \le \hat{\beta}_1 + t_{0.025,18}\sqrt{\frac{\hat{\sigma}^2}{S_{xx}}}$$

or

$$14.97 - 2.101\sqrt{\frac{1.17}{0.68}} \le \beta_1 \le 14.97 + 2.101\sqrt{\frac{1.17}{0.68}}$$

This simplifies to

$$12.21 \le \beta_1 \le 17.73$$

10-6.2 Confidence Interval on the Mean Response

A confidence interval may be constructed from the mean response at a specified value of x, say, x_0. This is a confidence interval about $E(Y \mid x_0) = \mu_{Y|x_0}$ and is often called a confidence interval about the regression line. Since $E(Y \mid x_0) = \mu_{Y|x_0} = \beta_0 + \beta_1 x_0$, we may obtain a point estimate of $\mu_{Y|x_0}$ from the fitted model as

$$\hat{\mu}_{Y|x_0} = \hat{\beta}_0 + \hat{\beta}_1 x_0$$

Now $\hat{\mu}_{Y|x_0}$ is an unbiased point estimator of $\mu_{Y|x_0}$, since $\hat{\beta}_0$ and $\hat{\beta}_1$ are unbiased estimators of β_0 and β_1. The variance of $\hat{\mu}_{Y|x_0}$ is

$$V(\hat{\mu}_{Y|x_0}) = \sigma^2\left[\frac{1}{n} + \frac{(x_0 - \bar{x})^2}{S_{xx}}\right]$$

This last result follows from the fact that $\text{cov}(\bar{Y}, \hat{\beta}_1) = 0$. (Refer to Exercise 10-71.) Also, $\hat{\mu}_{Y|x_0}$ is normally distributed, because $\hat{\beta}_1$ and $\hat{\beta}_0$ are normally distributed, and if we use $\hat{\sigma}^2$ as an estimate of σ^2, it is easy to show that

$$\frac{\hat{\mu}_{Y|x_0} - \mu_{Y|x_0}}{\sqrt{\hat{\sigma}^2\left[\dfrac{1}{n} + \dfrac{(x_0 - \bar{x})^2}{S_{xx}}\right]}}$$

has a t distribution with $n - 2$ degrees of freedom. This leads to the following confidence interval definition.

Definition

A $100(1 - \alpha)$ percent **confidence interval about the mean response** at the value of $x = x_0$, say $\mu_{Y|x_0}$, is given by

$$\hat{\mu}_{Y|x_0} - t_{\alpha/2,n-2}\sqrt{\hat{\sigma}^2\left[\frac{1}{n} + \frac{(x_0 - \bar{x})^2}{S_{xx}}\right]}$$
$$\leq \mu_{Y|x_0} \leq \hat{\mu}_{Y|x_0} + t_{\alpha/2,n-2}\sqrt{\hat{\sigma}^2\left[\frac{1}{n} + \frac{(x_0 - \bar{x})^2}{S_{xx}}\right]} \qquad (10\text{-}33)$$

where $\hat{\mu}_{Y|x_0} = \hat{\beta}_0 + \hat{\beta}_1 x_0$ is computed from the fitted regression model.

Note that the width of the confidence interval for $\mu_{Y|x_0}$ is a function of the value specified for x_0. The interval width is a minimum for $x_0 = \bar{x}$ and widens as $|x_0 - \bar{x}|$ increases.

EXAMPLE 10-6

We will construct a 95% confidence interval about the mean response for the data in Example 10-1. The fitted model is $\hat{\mu}_{Y|x_0} = 74.20 + 14.97x_0$, and the 95% confidence interval on $\mu_{Y|x_0}$ is found from equation 10-33 as

$$\hat{\mu}_{Y|x_0} \pm 2.101 \sqrt{1.17\left[\frac{1}{20} + \frac{(x_0 - 1.20)^2}{0.68}\right]}$$

Suppose that we are interested in predicting mean oxygen purity when $x_0 = 1.00\%$. Then

$$\hat{\mu}_{Y|x_0} = 74.20 + 14.97(1.00) = 89.17$$

and the 95% confidence interval is

$$\left[89.17 \pm 2.101 \sqrt{1.17\left[\frac{1}{20} + \frac{(1.00 - 1.20)^2}{0.68}\right]}\right]$$

or

$$89.17 \pm 0.75$$

Therefore, the 95% confidence interval on $\mu_{Y|1.00}$ is

$$88.42 \leq \mu_{Y|1.00} \leq 89.92$$

By repeating these calculations for several different values for x_0 we can obtain confidence limits for each corresponding value of $\mu_{Y|x_0}$. Figure 10-7 displays the scatter diagram with the fitted model and the corresponding 95% confidence limits plotted as the upper and lower lines. The 95% confidence level applies only to the interval obtained at one value of x, and not to the entire set of x-levels. Notice that the width of the confidence interval on $\mu_{Y|x_0}$ increases as $|x_0 - \bar{x}|$ increases.

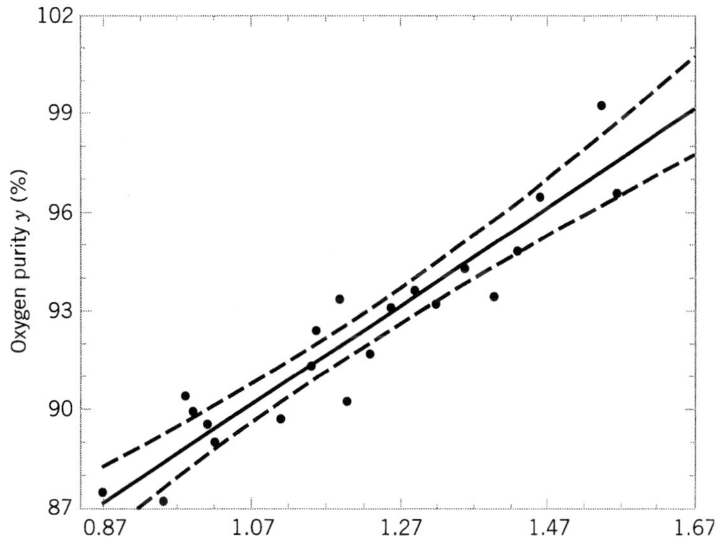

Figure 10-7 Scatter diagram of oxygen purity data from Example 10-1 with fitted regression line and 95 percent confidence limits on $\mu_{Y|x_0}$.

10-7 PREDICTION OF NEW OBSERVATIONS

An important application of a regression model is predicting new or future observations Y corresponding to a specified level of the regressor variable x. If x_0 is the value of the regressor variable of interest, then

$$\hat{Y}_0 = \hat{\beta}_0 + \hat{\beta}_1 x_0 \tag{10-34}$$

is the point estimator of the new or future value of the response Y_0.

Now consider obtaining an interval estimate for this future observation Y_0. This new observation is independent of the observations used to develop the regression model. Therefore, the confidence interval for $\mu_{Y|x_0}$ in equation 10-33 is inappropriate, since it is based only on the data used to fit the regression model. The confidence interval about $\mu_{Y|x_0}$ refers to the true mean response at $x = x_0$ (that is, a population parameter), not to future observations.

Let Y_0 be the future observations at $x = x_0$, and let $\hat{Y}_0$ given by equation 10-34 be the estimator of Y_0. Note that the error in prediction

$$\psi = Y_0 - \hat{Y}_0$$

is a normally distributed random variable with mean zero and variance

$$\begin{aligned} V(\psi) &= V(Y_0 - \hat{Y}_0) \\ &= \sigma^2\left[1 + \frac{1}{n} + \frac{(x_0 - \bar{x})^2}{S_{xx}}\right] \end{aligned}$$

because Y_0 is independent of $\hat{Y}_0$. If we use $\hat{\sigma}^2$ to estimate σ^2, we can show that

$$\frac{Y_0 - \hat{Y}_0}{\sqrt{\hat{\sigma}^2\left[1 + \frac{1}{n} + \frac{(x_0 - \bar{x})^2}{S_{xx}}\right]}}$$

has a t distribution with $n - 2$ degrees of freedom. From this we can develop the following **prediction interval** definition.

Definition

A $100(1 - \alpha)$ percent **prediction interval on a future observation** y_0 at the value x_0 is given by

$$\hat{y}_0 - t_{\alpha/2,n-2}\sqrt{\hat{\sigma}^2\left[1 + \frac{1}{n} + \frac{(x_0 - \bar{x})^2}{S_{xx}}\right]} \le y_0 \le \hat{y}_0 + t_{\alpha/2,n-2}\sqrt{\hat{\sigma}^2\left[1 + \frac{1}{n} + \frac{(x_0 - \bar{x})^2}{S_{xx}}\right]} \tag{10-35}$$

The value $\hat{y}_0$ is computed from the regression model $\hat{y}_0 = \hat{\beta}_0 + \hat{\beta}_1 x_0$.

Notice that the prediction interval is of minimum width at $x_0 = \bar{x}$ and widens as $|x_0 - \bar{x}|$ increases. By comparing equation 10-35 with equation 10-33, we observe that the prediction interval at the point x_0 is always wider than the confidence interval at x_0. This results because the prediction interval depends on both the error from the fitted model and the error associated with future observations.

EXAMPLE 10-7

To illustrate the construction of a prediction interval, suppose we use the data in Example 10-1 and find a 95% prediction interval on the next observation of oxygen purity at $x_0 = 1.00\%$. Using equation 10-35 and recalling from Example 10-6 that $\hat{y}_0 = 89.17$, we find that the prediction interval is

$$89.17 - 2.101\sqrt{1.17\left[1 + \frac{1}{20} + \frac{(1.00 - 1.20)^2}{0.68}\right]} \leq y_0 \leq 89.17 + 2.101\sqrt{1.17\left[1 + \frac{1}{20} + \frac{(1.00 - 1.20)^2}{0.68}\right]}$$

which simplifies to

$$86.78 \leq y_0 \leq 91.56$$

By repeating the foregoing calculations at different levels of x_0, we may obtain the 95% prediction intervals shown graphically as the lower and upper lines about the fitted regression model in Fig. 10-8. Notice that this graph also shows the 95% confidence limits on $\mu_{Y|x_0}$ calculated in Example 10-6. It illustrates that the prediction limits are always wider than the confidence limits.

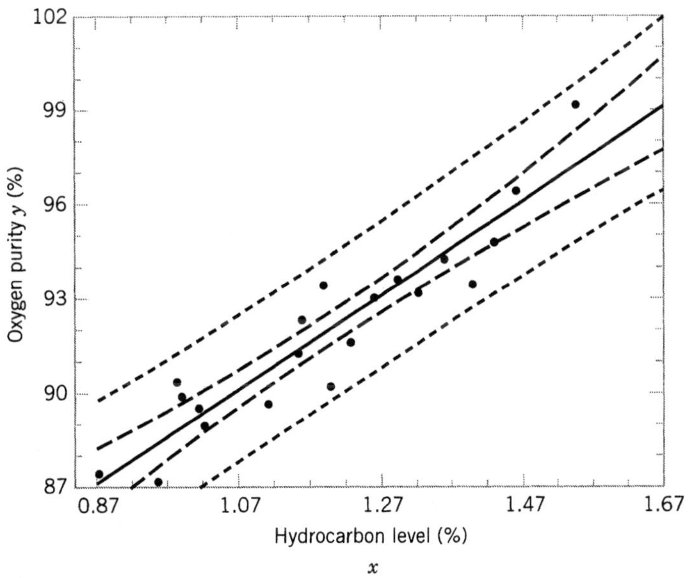

Figure 10-8 Scatter diagram of oxygen purity data from Example 10-1 with fitted regression line, 95% prediction limits (outer lines) and 95% confidence limits on $\mu_{Y|x_0}$.

EXERCISES FOR SECTIONS 10-6 AND 10-7

10-31. Refer to the data in Exercise 10-1 on y = intrinsic permeability of concrete and x = compressive strength. Find a 95% confidence interval on each of the following:
(a) slope
(b) intercept
(c) mean permeability when $x = 2.5$
(d) Find a 95% prediction interval on permeability when $x = 2.5$. Explain why this interval is wider than the interval in part (c).

10-32. Exercise 10-2 presented data on roadway surface temperature x and pavement deflection y. Find a 99% confidence interval on each of the following:
(a) slope
(b) intercept
(c) mean deflection when temperature $x = 85$°F
(d) Find a 99% prediction interval on pavement deflection when the temperature is 90°F.

10-33. Exercise 10-4 presented data on the number of games won by NFL teams in 1976. Find a 95% confidence interval on each of the following:
(a) slope
(b) intercept
(c) mean number of games won when opponents rushing yardage is limited to $x = 1800$
(d) Find a 95% prediction interval on the number of games won when opponents rushing yards is 1800.

10-34. Refer to the data on y = house selling price and x = taxes paid in Exercise 10-5. Find a 95% confidence interval on each of the following:
(a) β_1
(b) β_0
(c) mean selling price when the taxes paid are $x = 7.50$
(d) Compute the 95% prediction interval for selling price when the taxes paid are $x = 7.50$.

10-35. Exercise 10-6 presented data on y = steam usage and x = monthly average temperature.
(a) Find a 99% confidence interval for β_1.
(b) Find a 99% confidence interval for β_0.
(c) Find a 95% confidence interval on mean steam usage when the average temperature is 55°F.
(d) Find a 95% prediction interval on steam usage when temperature is 55°F. Explain why this interval is wider than the interval in part (c).

10-36. Exercise 10-7 presented gasoline mileage performance for 20 cars, along with information about the engine displacement. Find a 95% confidence interval on each of the following:
(a) slope
(b) intercept
(c) mean highway gasoline mileage when the engine displacement is $x = 150$ in^3
(d) Construct a 95% prediction interval on highway gasoline mileage when the engine displacement is $x = 150$ in^3.

10-37. Consider the data in Exercise 10-8 on y = green liquor Na_2S concentration and x = production in a paper mill. Find a 99% confidence interval on each of the following:
(a) β_1
(b) β_0
(c) mean Na_2S concentration when production $x = 910$ tons/day
(d) Find a 99% prediction interval on Na_2S concentration when $x = 910$ tons/day.

10-38. Exercise 10-9 presented data on y = blood pressure rise and x = sound pressure level. Find a 95% confidence interval on each of the following:
(a) β_1
(b) β_0

(c) mean blood pressure rise when the sound pressure level is 85 dB
(d) Find a 95% prediction interval on blood pressure rise when the sound pressure level is 85 dB.

10-39. Refer to the data in Exercise 10-10 on y = wear volume of mild steel and x = oil viscosity. Find a 95% confidence interval on each of the following:
(a) intercept
(b) slope
(c) mean wear when oil viscosity $x = 30$

10-40. Exercise 10-11 presented data on chloride concentration y and roadway area x on watersheds in central Rhode Island. Find a 99% confidence interval on each of the following:
(a) β_1
(b) β_0
(c) mean chloride concentration when roadway area $x = 1.0\%$
(d) Find a 99% prediction interval on chloride concentration when roadway area $x = 1.0\%$.

10-41. Refer to the data in Exercise 10-12 on rocket motor shear strength y and propellant age x. Find a 95% confidence interval on each of the following:
(a) slope β_1
(b) intercept β_0
(c) mean shear strength when age x = 20 weeks
(d) Find a 95% prediction interval on shear strength when age x = 20 weeks.

10-8 ASSESSING THE ADEQUACY OF THE REGRESSION MODEL

Fitting a regression model requires several assumptions. Estimation of the model parameters requires the assumption that the errors are uncorrelated random variables with mean zero and constant variance. Tests of hypotheses and interval estimation require that the errors be normally distributed. In addition, we assume that the order of the model is correct; that is, if we fit a simple linear regression model, then we are assuming that the phenomenon actually behaves in a linear or first-order manner.

The analyst should always consider the validity of these assumptions to be doubtful and conduct analyses to examine the adequacy of the model that has been tentatively entertained. In this section we discuss methods useful in this respect.

10-8.1 Residual Analysis

The residuals from a regression model are $e_i = y_i - \hat{y}_i$, $i = 1, 2, \ldots, n$, where y_i is an actual observation and $\hat{y}_i$ is the corresponding fitted value from the regression model. Analysis of the residuals is frequently helpful in checking the assumption that the errors are approximately normally distributed with constant variance, as well as in determining whether additional terms in the model would be useful.

As an approximate check of normality, the experimenter can construct a frequency histogram of the residuals or a **normal probability plot of residuals.** Many computer programs will produce a normal probability plot of residuals, and since the sample sizes in regression are often too small for a histogram to be meaningful, the normal probability plotting method is preferred. It requires judgment to assess the abnormality of such plots. (Refer to the discussion of the "fat pencil" method in Chapter 5).

We may also *standardize* the residuals by computing $d_i = e_i/\sqrt{\hat{\sigma}^2}$, $i = 1, 2, \ldots, n$. If the errors are normally distributed, then approximately 95% of the standardized residuals should fall in the interval $(-2, +2)$. Residuals that are far outside this interval may indicate the presence of an **outlier**, that is, an observation that is not typical of the rest of the data. Various rules have been proposed for discarding outliers. However, outliers sometimes provide important information about unusual circumstances of interest to experimenters and should not be discarded. For further discussion of outliers, see Montgomery and Peck (1992).

It is frequently helpful to plot the residuals (1) in time sequence (if known), (2), against the $\hat{y}_i$, and (3) against the independent variable x. These graphs will usually look like one of the four general patterns shown in Fig. 10-9. Pattern (*a*) in Fig. 10-9 represents the ideal situation, while patterns (*b*), (*c*), and (*d*) represent anomalies. If the residuals appear as in (*b*), the variance of the observations may be increasing with time or with the magnitude of y_i or x_i. Data transformation on the response y is often used to eliminate this problem. Widely used variance-stabilizing transformations include the use of $\sqrt{y}$, $ln\ y$, or $1/y$ as the response. See Montgomery and Peck (1992) for more details regarding methods for selecting an appropriate transformation. If a plot of the residuals against time

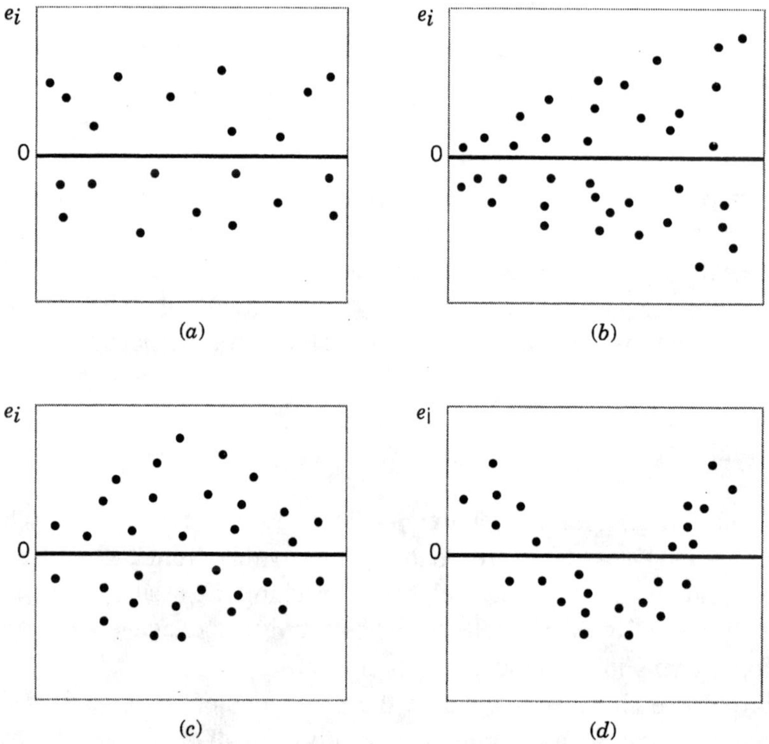

Figure 10-9 Patterns for residual plots. (a) satisfactory, (b) funnel, (c) double bow, (d) nonlinear. [Adapted from Montgomery and Peck (1992).]

has the appearance of (*b*), then the variance of the observations is increasing with time. Plots of residuals against $\hat{y}_i$ and x_i that look like (*c*) also indicate inequality of variance. Residual plots that look like (*d*) indicate model inadequacy; that is, higher order terms should be added to the model, a transformation on the x-variable or the y-variable (or both) should be considered, or other regressors should be considered.

EXAMPLE 10-8

The regression model for the oxygen purity data in Example 10-1 is $\hat{y} = 74.20 + 14.97x$. Table 10-5 presents the observed and predicted values of y at each value of x from this data set, along with the corresponding residual. These values were computed using Minitab and show the number of decimal places typical of computer output. A normal probability plot of the residuals is shown in Fig. 10-10. Since the residuals fall approximately along a straight line in the figure, we conclude that there is no severe departure from normality. The residuals are also plotted against the predicted value $\hat{y}_i$ in Fig. 10-11 and against the hydrocarbon levels x_i in Fig. 10-12. These plots do not indicate any serious model inadequacies.

Table 10-5 Oxygen Purity Data from Example 10-1, Predicted Values, and Residuals

	Hydrocarbon Level, x	Oxygen Purity, y	Predicted Value, $\hat{y}$	Residual $e = y - \hat{y}$
1	0.99	90.01	89.069009	0.940991
2	1.02	89.05	89.518136	−0.468136
3	1.15	91.43	91.464353	−0.034353
4	1.29	93.74	93.560279	0.179721
5	1.46	96.73	96.105332	0.624668
6	1.36	94.45	94.608242	−0.158242
7	0.87	87.59	87.272501	0.317499
8	1.23	91.77	92.662025	−0.892025
9	1.55	99.42	97.452713	1.967287
10	1.40	93.65	95.207078	−1.557078
11	1.19	93.54	92.063189	1.476811
12	1.15	92.52	91.614062	0.905938
13	0.98	90.56	88.919300	1.640700
14	1.01	89.54	89.368427	0.171573
15	1.11	89.85	90.865517	−1.015517
16	1.20	90.39	92.212898	−1.822898
17	1.26	93.25	93.111152	0.138848
18	1.32	93.41	94.009406	−0.599406
19	1.43	94.98	95.656205	−0.676205
20	0.95	87.33	88.470173	−1.140173

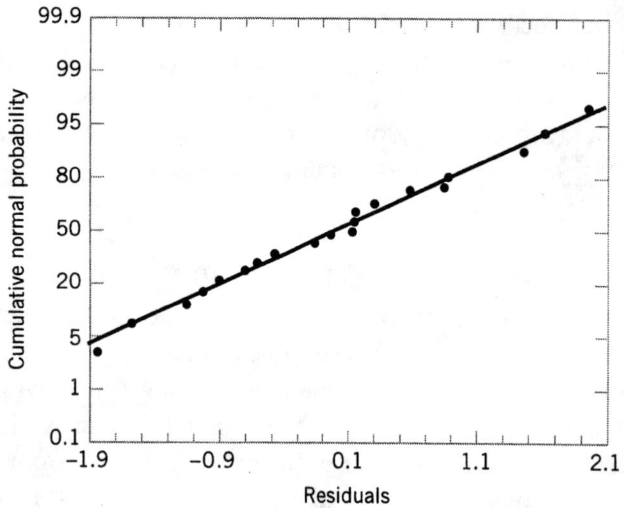

Figure 10-10 Normal probability plot of residuals, Example 10-8.

10-8.2 Coefficient of Determination (R^2)

The quantity

$$R^2 = \frac{SS_R}{SS_T} = 1 - \frac{SS_E}{SS_T} \tag{10-36}$$

is called the **coefficient of determination** and is often used to judge the adequacy of a regression model. Subsequently, we will see that in the case where X and Y are jointly distributed random variables, R^2 is the square of the correlation coefficient between X and Y. From the analysis of variance identity in equation 10-25 and 10-26, $0 \leq R^2 \leq 1$.

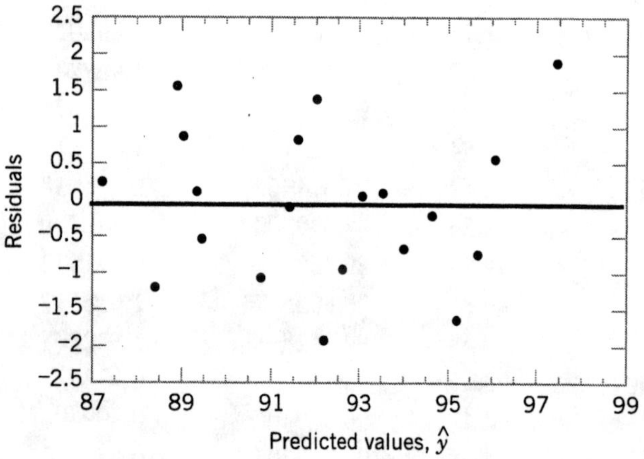

Figure 10-11 Plot of residuals versus predicted oxygen purity $\hat{y}$, Example 10-8.

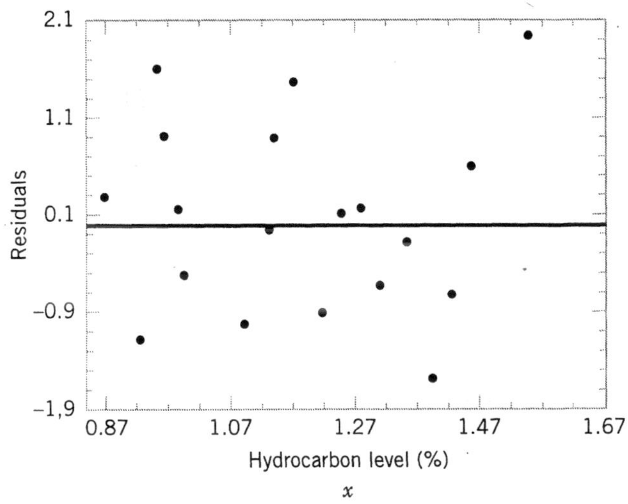

Figure 10-12 Plot of residuals versus hydrocarbon level x, Example 10-8.

We often refer loosely to R^2 as the amount of variability in the data explained or accounted for by the regression model. For the oxygen purity regression model, we have $R^2 = SS_R/SS_T = 152.39/173.37 = 0.8790$; that is, the model accounts for 87.90% of the variability in the data.

The statistic R^2 should be used with caution, because it is always possible to make R^2 unity by simply adding enough terms to the model. For example, we can obtain a "perfect" fit to n data points with a polynomial of degree $n - 1$. In addition, R^2 will always increase if we add a variable to the model, but this does not necessarily imply that the new model is superior to the old one. Unless the error sum of squares in the new model is reduced by an amount equal to the original error mean square, the new model will have a larger error mean square than the old one, because of the loss of one error degree of freedom. Thus, the new model will actually be worse than the old one.

There are several misconceptions about R^2. In general, R^2 does not measure the magnitude of the slope of the regression line. A large value of R^2 does not imply a steep slope. Furthermore, R^2 does not measure the appropriateness of the model, since it can be artificially inflated by adding higher order polynomial terms in x to the model. Even if y and x are related in a nonlinear fashion, R^2 will often be large. For example, R^2 for the regression equation in Fig. 10-6(b) will be relatively large, even though the linear approximation is poor. Finally, even though R^2 is large, this does not necessarily imply that the regression model will provide accurate predictions of future observations.

10-8.3 Lack-of-Fit Test

Regression models are often fit to data to provide an empirical model when the true relationship between the variables Y and x is unknown. Naturally, we would like to know whether the order of the model tentatively assumed is correct. This section describes a test for the validity of this assumption.

The danger of using a regression model that is a poor approximation of the true functional relationship is illustrated in Fig. 10-13. Obviously, a polynomial of degree two or greater in x should have been used in this situation.

We present a test for the "goodness of fit" of the regression model. Specifically, the hypotheses we wish to test are

H_0: The simple linear regression model is correct.

H_1: The simple linear regression model is not correct.

The test involves partitioning the error or residual sum of squares into the following components:

$$SS_E = SS_{PE} + SS_{LOF} \tag{10-37}$$

where SS_{PE} is the sum of squares attributable to **pure error,** and SS_{LOF} is the sum of squares attributable to the **lack-of-fit** of the model. To compute SS_{PE}, we must have repeated observations on the response Y for at least one level of x. Suppose we have n total observations such that

$y_{11}, y_{12}, \ldots, y_{1n_1}$ repeated observations at x_1

$y_{21}, y_{22}, \ldots, y_{2n_2}$ repeated observations at x_2

$\vdots$

$y_{m1}, y_{m2}, \ldots, y_{mn_m}$ repeated observations at x_m

Note that there are m distinct levels of x. The contribution to the pure-error sum of squares at x_1 (say) would be

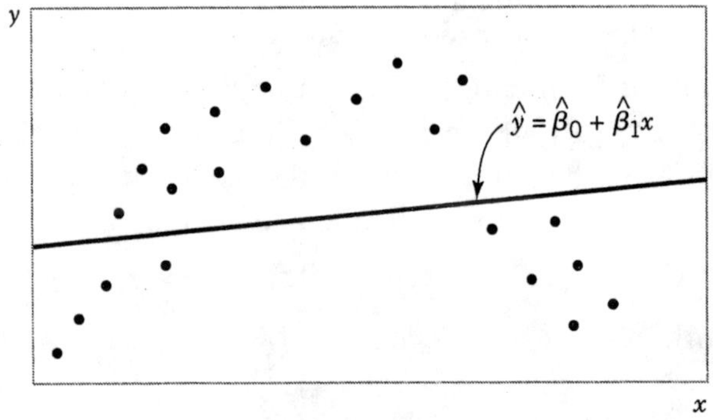

Figure 10-13 A regression model displaying lack of fit.

$$\sum_{u=1}^{n_1} (y_{1u} - \bar{y}_1)^2 \tag{10-38}$$

where $\bar{y}_1$ represents the average of all n_1 repeat observations on the response y at x_1. The total sum of squares for pure error would be obtained by summing equation 10-38 over all levels of x as

$$SS_{PE} = \sum_{i=1}^{m} \sum_{u=1}^{n_i} (y_{iu} - \bar{y}_i)^2 \tag{10-39}$$

There are $n_{pe} = \sum_{i=1}^{m} (n_i - 1) = n - m$ degrees of freedom associated with the pure-error sum of squares. The sum of squares for lack of fit is simply

$$SS_{LOF} = SS_E - SS_{PE} \tag{10-40}$$

with $n - 2 - n_{pe} = m - 2$ degrees of freedom. The test statistic for lack of fit would then be

$$F_0 = \frac{SS_{LOF}/(m - 2)}{SS_{PE}/(n - m)} = \frac{MS_{LOF}}{MS_{PE}} \tag{10-41}$$

and we would reject the hypothesis that the model adequately fits the data if $f_0 > f_{\alpha, m-2, n-m}$.

This test procedure may be easily introduced into the analysis of variance conducted for the significance of regression. If the null hypothesis of model adequacy is rejected, then the model must be abandoned and attempts must be made to find a more appropriate model. If H_0 is not rejected, then there is no apparent reason to doubt the adequacy of the model, and MS_{PE} and MS_{LOF} are often combined to estimate σ^2.

EXAMPLE 10-9

Consider the data on two variables y and x shown below. Fit a simple linear regression model and test for lack of fit, using $\alpha = 0.05$.

x	y
1.0	2.3, 1.8
2.0	2.8
3.3	1.8, 3.7
4.0	2.6, 2.6, 2.2
5.0	2.0

x	y
5.6	3.5, 2.8, 2.1
6.0	3.4, 3.2
6.5	3.4
6.9	5.0

Table 10-6 Analysis of Variance for Example 10-9

Source of Variation	Sum of Squares	Degrees of Freedom	Mean Square	f_0	P-value
Regression	3.4930	1	3.4930	6.66	0.0218
Residual	7.3372	14	0.5241		
(Lack of fit)	4.3005	7	0.6144	1.42	0.3276
(Pure error)	3.0366	7	0.4338		
Total	10.8300	15			

The regression model is $\hat{y} = 1.697 + 0.259x$, and the regression sum of squares is $SS_R = 3.4930$. The pure-error sum of squares is computed as follows:

Level of x	$\sum_{u=1}^{n_i} (y_{iu} - \bar{y}_i)^2$	Degrees of Freedom
1.0	0.1250	1
3.3	1.8050	1
4.0	0.0166	2
5.6	0.9800	2
6.0	0.0200	1
Totals	3.0366	7

The analysis of variance is summarized in Table 10-6. Since the lack-of-fit F-statistic is $f_0 = 1.42$, which has a P-value of $P = 0.3276$, we cannot reject the hypothesis that the tentative model adequately describes the data. We will pool lack-of-fit and pure-error mean squares to form the residual mean square that is the denominator mean square in the test for significance of regression. In addition, since the P-value for the statistic $f_0 = 6.66$ with 1 and 14 degrees of freedom associated with significance of regression is $P = 0.0216$, we conclude that $\beta_1 \neq 0$.

In fitting a regression model to experimental data, a good practice is to use the lowest degree model that adequately describes the data. The lack-of-fit test may be useful in this respect. However, it is always possible to fit a polynomial of degree $n - 1$ to n data points, and the experimenter should not consider using a model that is "saturated"—that is, that has very nearly as many independent variables as observations on y.

EXERCISES FOR SECTION 10-8

10-42. Refer to the NFL team performance data in Exercise 10-4.

(a) Calculate R^2 for this model and provide a practical interpretation of this quantity.

(b) Prepare a normal probability plot of the residuals from the least squares model. Does the normality assumption seem to be satisfied?

(c) Plot the residuals versus $\hat{y}$ and against x. Interpret these graphs.

10-43. Refer to the data in Exercise 10-5 on house selling price y and taxes paid x.

(a) Find the residuals for the least squares model.

(b) Prepare a normal probability plot of the residuals and interpret this display.

(c) Plot the residuals versus $\hat{y}$ and versus x. Does the assumption of constant variance seem to be satisfied?

(d) What proportion of total variability is explained by the regression model?

10-44. Exercise 10-6 presents data on y = steam usage and x = average monthly temperature.

(a) What proportion of total variability is accounted for by the simple linear regression model?

(b) Prepare a normal probability plot of the residuals and interpret this graph.

(c) Plot residuals versus $\hat{y}$ and x. Do the regression assumptions appear to be satisfied?

10-45. Refer to the gasoline mileage data in Exercise 10-7.

(a) What proportion of total variability in highway gasoline mileage performance is accounted for by engine displacement?

(b) Plot the residuals versus $\hat{y}$ and x, and comment on the graphs.

(c) Prepare a normal probability plot of the residuals. Does the normality assumption appear to be satisfied?

10-46. Consider the data in Exercise 10-8 on y = green liquor Na_2S concentration and x = paper machine production. Suppose that a 14th sample point is added to the original data, where $y_{14} = 59$ and $x_{14} = 855$.

(a) Prepare a scatter diagram of y versus x. Fit the simple linear regression model to all 14 observations.

(b) Test for significance of regression with $\alpha = 0.05$.

(c) Estimate σ^2 for this model.

(d) Compare the estimate of σ^2 obtained in part (c) above with the estimate of σ^2 obtained from the original 13 points. Which estimate is larger and why?

(e) Compute the residuals for this model. Does the value of e_{14} appear unusual?

(f) Prepare and interpret a normal probability plot of the residuals.

(g) Plot the residuals versus $\hat{y}$ and versus x. Comment on these graphs.

10-47. Refer to Exercise 10-9, which presented data on blood pressure rise y and sound pressure level x.

(a) What proportion of total variability in blood pressure rise is accounted for by sound pressure level?

(b) Test for lack of fit using $\alpha = 0.05$. What is the P-value for this test?

(c) Prepare a normal probability plot of the residuals from this least squares model. Interpret this plot.

(d) Plot residuals versus $\hat{y}$ and versus x. Comment on these plots.

10-48. Exercise 10-10 presents data on wear volume y and oil viscosity x.

(a) Calculate R^2 for this model. Provide an interpretation of this quantity.

(b) Plot the residuals from this model versus $\hat{y}$ and versus x. Interpret these plots.

(c) Prepare a normal probability plot of the residuals. Does the normality assumption appear to be satisfied?

10-49. Refer to Exercise 10-11 which presented data on chloride concentration y and roadway area x.

(a) What proportion of the total variability in chloride concentration is accounted for by the regression model?

(b) Use the repeat observations at $x = 70$ and $x = 78$ to obtain a 2-degree-of-freedom estimate of pure error.

(c) Use the pure-error quantity computed in part (b) to test for lack of fit of the regression model. Use $\alpha = 0.05$. What conclusion can you draw about model adequacy?

(d) Plot the residuals versus $\hat{y}$ and versus x. Interpret these plots.

(e) Prepare a normal probability plot of the residuals. Does the normality assumption appear to be satisfied?

10-50. Consider the rocket propellant data in Exercise 10-12.

(a) Calculate R^2 for this model. Provide an interpretation of this quantity.

(b) Plot the residuals on a normal probability scale. Do any points seem unusual on this plot?

(c) Delete the two points identified in part (b) from the sample and fit the simple linear regression model to the remaining 18 points. Calculate the value of R^2 for the new model. Is it larger or smaller than the value of R^2 computed in part (a)? Why?

(d) Did the value of $\hat{\sigma}^2$ change dramatically when the two points identified above were deleted and the model fit to the remaining points? Why?

10-51. Show that an equivalent way to define the test for significance of regression in simple linear regression is to base the test on R^2 as follows: to test H_0: $\beta_1 = 0$ versus H_1: $\beta_1 \neq 0$, calculate

$$F_0 = \frac{R^2(n-2)}{1-R^2}$$

and to reject H_0: $\beta_1 = 0$ if the computed value $f_0 > f_{\alpha,1,n-2}$.

10-52. Suppose that the simple linear regression model has been fit to $n = 25$ observations and $R^2 = 0.90$.

(a) Test for significance of regression at $\alpha = 0.05$. Use the results of Exercise 10-51.

(b) What is the smallest value of R^2 that would lead to the conclusion of a significant regression if $\alpha = 0.05$?

10-53. Consider the rocket propellant data in Exercise 10-12. Calculate the standardized residuals for these data. Does this provide any helpful information about the magnitude of the residuals?

10-54. **Studentized Residuals.** Show that the variance of the ith residual is

$$V(e_i) = \sigma^2\left[1 - \left(\frac{1}{n} + \frac{(x_i - \bar{x})^2}{S_{xx}}\right)\right]$$

Hint:

$$\text{cov}(Y_i, \hat{Y}_i) = \sigma^2\left[\frac{1}{n} + \frac{(x_i - \bar{x})^2}{S_{xx}}\right].$$

The ith studentized residual is defined as

$$r_i = \frac{e_i}{\sqrt{\hat{\sigma}^2\left[1 - \left(\frac{1}{n} + \frac{(x_i - \bar{x})^2}{S_{xx}}\right)\right]}}$$

(a) Explain why r_i has unit standard deviation.

(b) Do the standardized residuals have unit standard deviation?

(c) Discuss the behavior of the standardized residual when the sample value x_i is very close to the middle of the range of x.

(d) Discuss the behavior of the studentized residual when the sample value x_i is very near one end of the range of x.

10-9 TRANSFORMATIONS TO A STRAIGHT LINE

We occasionally find that the straight-line regression model $Y = \beta_0 + \beta_1 x + \epsilon$ is inappropriate because the true regression function is nonlinear. Sometimes nonlinearity is visually determined from the scatter diagram, and sometimes, because of prior experience or underlying theory, we know in advance that the model is nonlinear. Occasionally, a scatter diagram will exhibit an apparent nonlinear relationship between Y and x. In some of these situations, a nonlinear function can be expressed as a straight line by using a suitable transformation. Such nonlinear models are called **intrinsically linear**.

As an example of a nonlinear model that is intrinsically linear, consider the exponential function

$$Y = \beta_0 e^{\beta_1 x} \epsilon$$

This function is intrinsically linear, since it can be transformed to a straight line by a logarithmic transformation

$$\ln Y = \ln \beta_0 + \beta_1 x + \ln \epsilon$$

This transformation requires that the transformed error terms $\ln \epsilon$ are normally and independently distributed with mean 0 and variance σ^2.

Another intrinsically linear function is

$$Y = \beta_0 + \beta_1 \left(\frac{1}{x}\right) + \epsilon$$

By using the reciprocal transformation $z = 1/x$, the model is linearized to

$$Y = \beta_0 + \beta_1 z + \epsilon$$

Sometimes several transformations can be employed jointly to linearize a function. For example, consider the function

$$Y = \frac{1}{\exp(\beta_0 + \beta_1 x + \epsilon)}$$

letting $Y^* = 1/Y$, we have the linearized form

$$\ln Y^* = \beta_0 + \beta_1 x + \epsilon$$

Daniel and Wood (1980) give several nonlinear models that are intrinsically linear. For examples of fitting those models, refer to Montgomery and Peck (1992) or Myers (1990).

10-10 CORRELATION

Our development of regression analysis has assumed that x is a mathematical variable, measured with negligible error, and that Y is a random variable. Many applications of regression analysis involve situations in which both X and Y are random variables. In these situations, it is usually assumed that the observations (X_i, Y_i), $i = 1, 2, \ldots, n$ are jointly distributed random variables obtained from the distribution $f(x, y)$.

For example, suppose we wish to develop a regression model relating the shear strength of spot welds to the weld diameter. In this example, weld diameter cannot be controlled. We would randomly select n spot welds and observe a diameter (X_i) and a shear strength (Y_i) for each. Therefore (X_i, Y_i) are jointly distributed random variables.

We assume that the joint distribution of X_i and Y_i is the bivariate normal distribution presented in Chapter 6, and μ_Y and σ_Y^2 are the mean and variance of Y, μ_X and σ_X^2 are the mean and variance of X, and ρ is the correlation coefficient between Y and X. Recall that the correlation coefficient is defined as

$$\rho = \frac{\sigma_{XY}}{\sigma_X \sigma_Y} \tag{10-42}$$

where σ_{XY} is the covariance between Y and X.

The conditional distribution of Y for a given value of $X = x$ is

$$f_{Y|x}(y) = \frac{1}{\sqrt{2\pi}\sigma_{Y|x}} \exp\left[-\frac{1}{2}\left(\frac{y - \beta_0 - \beta_1 x}{\sigma_{Y|x}}\right)^2\right] \tag{10-43}$$

where

$$\beta_0 = \mu_Y - \mu_X \rho \frac{\sigma_Y}{\sigma_X} \tag{10-44}$$

$$\beta_1 = \frac{\sigma_Y}{\sigma_X}\rho \tag{10-45}$$

and the variance of the conditional distribution of Y given $X = x$ is

$$\sigma_{Y|x}^2 = \sigma_Y^2(1 - \rho^2) \tag{10-46}$$

That is, the conditional distribution of Y given $X = x$ is normal with mean

$$E(Y\,|\,x) = \beta_0 + \beta_1 x \tag{10-47}$$

and variance $\sigma_{Y|x}^2$. Thus, the mean of the conditional distribution of Y given $X = x$ is a simple linear regression model. Furthermore, there is a relationship between the correlation coefficient ρ and the slope β_1. From equation 10-45 we see that if $\rho = 0$, then $\beta_1 = 0$, which implies that there is no regression of Y on X. That is, knowledge of X does not assist us in predicting Y.

The method of maximum likelihood may be used to estimate the parameter β_0 and β_1. It can be shown that the maximum likelihood estimators of those parameters are

$$\hat{\beta}_0 = \overline{Y} - \hat{\beta}_1 \overline{X} \tag{10-48}$$

and

$$\hat{\beta}_1 = \frac{\sum_{i=1}^{n} Y_i(X_i - \overline{X})}{\sum_{i=1}^{n} (X_i - \overline{X})^2} = \frac{S_{XY}}{S_{XX}} \tag{10-49}$$

We note that the estimators of the intercept and slope in equations 10-48 and 10-49 are identical to those given by the method of least squares in the case where X was assumed to be a mathematical variable. That is, the regression model with Y and X jointly normally distributed is equivalent to the model with X considered as a mathematical variable. This follows because the random variables Y given $X = x$ are independently and normally distributed with mean $\beta_0 + \beta_1 x$ and constant variance $\sigma^2_{Y|x}$. These results will also hold for any joint distribution of Y and X such that the conditional distribution of Y given X is normal.

It is possible to draw inferences about the correlation coefficient ρ in this model. The estimator of ρ is the **sample correlation coefficient**

$$R = \frac{\sum_{i=1}^{n} Y_i(X_i - \overline{X})}{\left[\sum_{i=1}^{n} (X_i - \overline{X})^2 \sum_{i=1}^{n} (Y_i - \overline{Y})^2\right]^{1/2}} = \frac{S_{XY}}{[S_{XX}S_{YY}]^{1/2}} \tag{10-50}$$

Note that

$$\hat{\beta}_1 = \left(\frac{S_{YY}}{S_{XX}}\right)^{1/2} R \tag{10-51}$$

so that the slope $\hat{\beta}_1$ is just the sample correlation coefficient R multiplied by a scale factor that is the square root of the "spread" of the Y values divided by the "spread" of the X values. Thus, $\hat{\beta}_1$ and R are closely related, although they provide somewhat different information. The sample correlation coefficient R measures the linear association between Y and X, while $\hat{\beta}_1$ measures the predicted change in the mean of Y for a unit change in

X. In the case of a mathematical variable x, R has no meaning because the magnitude of R depends on the choice of spacing of x. We may also write, from equation 10-51,

$$\begin{aligned} R^2 &= \hat{\beta}_1^2 \frac{S_{XX}}{S_{YY}} \\ &= \frac{\hat{\beta}_1 S_{XY}}{S_{YY}} \\ &= \frac{SS_R}{S_{YY}} \end{aligned}$$

which we recognize from equation 10-36 as the coefficient of determination. That is, the coefficient of determination R^2 is just the square of the correlation coefficient between Y and X.

It is often useful to test the hypotheses

$$\begin{aligned} H_0&: \rho = 0 \\ H_1&: \rho \neq 0 \end{aligned} \tag{10-52}$$

The appropriate test statistic for these hypotheses is

$$T_0 = \frac{R\sqrt{n-2}}{\sqrt{1-R^2}} \tag{10-53}$$

which has the t distribution with $n - 2$ degrees of freedom if H_0: $\rho = 0$ is true. Therefore, we would reject the null hypothesis if $|t_0| > t_{\alpha/2,n-2}$. This test is equivalent to the test of the hypothesis H_0: $\beta_1 = 0$ given in Section 10-5.1. This equivalence follows directly from equation 10-51.

The test procedure for the hypothesis

$$\begin{aligned} H_0&: \rho = \rho_0 \\ H_1&: \rho \neq \rho_0 \end{aligned} \tag{10-54}$$

where $\rho_0 \neq 0$ is somewhat more complicated. For moderately large samples (say, $n \geq 25$) the statistic

$$Z = \text{arctanh}\, R = \frac{1}{2} ln \frac{1+R}{1-R} \tag{10-55}$$

is approximately normally distributed with mean

$$\mu_Z = \text{arctanh}\, \rho = \frac{1}{2} \ln \frac{1+\rho}{1-\rho}$$

and variance

$$\sigma_Z^2 = (n-3)^{-1}$$

Therefore, to test the hypothesis H_0: $\rho = \rho_0$, we may use the test statistic

$$Z_0 = (\text{arctanh}\, R - \text{arctanh}\, \rho_0)(n-3)^{1/2} \qquad (10\text{-}56)$$

and reject H_0: $\rho = \rho_0$ if the value of the test statistic in equation 10-56 is such that $|z_0| > z_{\alpha/2}$.

It is also possible to construct an approximate $100(1-\alpha)$ percent confidence interval for ρ, using the transformation in equation 10-55. The approximate $100(1-\alpha)$ percent confidence interval is

$$\tanh\left(\text{arctanh}\, r - \frac{z_{\alpha/2}}{\sqrt{n-3}}\right) \le \rho \le \tanh\left(\text{arctanh}\, r + \frac{z_{\alpha/2}}{\sqrt{n-3}}\right) \qquad (10\text{-}57)$$

where $\tanh u = (e^u - e^{-u})/(e^u + e^{-u})$.

EXAMPLE 10-10

In Chapter 1 (Section 1-4) is described an application of regression analysis in which an engineer at a semiconductor assembly plant is investigating the relationship between pull strength of a wire bond and two factors: wire length and die height. In this example, we will consider only one of the factors, the wire length. A random sample of 25 units is selected and tested, and the wire bond pull strength and wire length are observed for each unit. The data are shown in Table 1-2, page 15. We assume that pull strength and wire length are jointly normally distributed.

Using the data in Table 1-2, we may calculate

$$SS_T = 6105.9447 \quad S_{xx} = 698.5600 \quad \text{and} \quad S_{xy} = 2027.7132$$

The regression model is

$$\hat{y} = 5.1145 + 2.9027x$$

The sample correlation coefficient between X and Y is computed from equation 10-50 as

$$r = \frac{S_{xy}}{[S_{xx}S_{yy}]^{1/2}} = \frac{2027.7132}{[(698.560)(6105.9447)]^{1/2}} = 0.9818$$

Note that $r^2 = (0.9818)^2 = 0.9640$, or that approximately 96.40% of the variability in pull strength is explained by the linear relationship to wire length.

Now suppose that we wish to test the hypothesis

$$H_0: \rho = 0$$
$$H_1: \rho \neq 0$$

with $\alpha = 0.05$. We can compute the t-statistic of equation 10-53 as

$$t_0 = \frac{r\sqrt{n-2}}{\sqrt{1-r^2}} = \frac{0.9818\sqrt{23}}{\sqrt{1-0.9640}} = 24.82$$

and since $t_{0.025,23} = 2.069$, we reject H_0 and conclude that the correlation coefficient $\rho \neq 0$.

Finally, we may construct an approximate 95% confidence interval on ρ from equation 10-57. Since arctanh r = arctanh 0.9818 = 2.3452, equation 10-57 becomes

$$\tanh\left(2.3452 - \frac{1.96}{\sqrt{22}}\right) \leq \rho \leq \tanh\left(2.3452 + \frac{1.96}{\sqrt{22}}\right)$$

which reduces to

$$0.9585 \leq \rho \leq 0.9921$$

EXERCISES FOR SECTION 10-10

10-55. The final test and exam averages for 20 randomly selected students taking a course in engineering statistics and a course in operations research are shown below. Assume that the final averages are jointly normally distributed.

(a) Find the regression line relating the statistics final average to the OR final average.

(b) Test for significance of regression using $\alpha = 0.05$.

Statistics	86	75	69	75	90
OR	80	81	75	81	92

Statistics	94	83	86	71	65
OR	95	80	81	76	72

Statistics	84	71	62	90	83
OR	85	72	65	93	81

Statistics	75	71	76	84	97
OR	70	73	72	80	98

(c) Estimate the correlation coefficient.
(d) Test the hypothesis that $\rho = 0$, using $\alpha = 0.05$.
(e) Test the hypothesis that $\rho = 0.5$, using $\alpha = 0.05$.
(f) Construct a 95% confidence interval for the correlation coefficient.

10-56. The weight and systolic blood pressure of 26 randomly selected males in the age group 25 to 30 are shown in the following table. Assume that weight and blood pressure are jointly normally distributed.

Subject	Weight	Systolic BP
1	165	130
2	167	133
3	180	150
4	155	128
5	212	151
6	175	146
7	190	150
8	210	140
9	200	148
10	149	125
11	158	133
12	169	135
13	170	150
14	172	153
15	159	128
16	168	132
17	174	149
18	183	158
19	215	150
20	195	163
21	180	156
22	143	124
23	240	170
24	235	165
25	192	160
26	187	159

(a) Find a regression line relating systolic blood pressure to weight.
(b) Test for significance of regression using $\alpha = 0.05$.
(c) Estimate the correlation coefficient.
(d) Test the hypothesis that $\rho = 0$, using $\alpha = 0.05$.
(e) Test the hypothesis that $\rho = 0.6$, using $\alpha = 0.05$.
(f) Construct a 95% confidence interval for the correlation coefficient.

10-57. Consider the NFL data introduced in Exercise 10-4.
(a) Estimate the correlation coefficient between the number of games won and the yards rushing by the opponents.
(b) Test the hypothesis H_0: $\rho = 0$ versus H_1: $\rho \neq 0$ using $\alpha = 0.05$. What is the P-value for this test?
(c) Construct a 95% confidence interval for ρ.
(d) Test the hypothesis H_0: $\rho = -0.7$ versus H_1: $\rho \neq -0.7$ using $\alpha = 0.05$. Find the P-value for this test.

10-58. Show that the t-statistic in equation 10-53 for testing H_0: $\rho = 0$ is identical to the t-statistic for testing H_0: $\beta_1 = 0$.

10-59. A random sample of 50 observations was made on the diameter of spot welds and the corresponding weld shear strength.
(a) Given that $r = 0.62$, test the hypothesis that $\rho = 0$, using $\alpha = 0.01$. What is the P-value for this test?
(b) Find a 99% confidence interval for ρ.
(c) Based on the confidence interval in part (b), can you conclude that $\rho = 0.5$ at the 0.01 level of significance?

10-60. Suppose that a random sample of 10,000 (X, Y) pairs yielded a sample correlation coefficient of $r = 0.02$.
(a) What is the conclusion that you would reach if you tested H_0: $\rho = 0$ using $\alpha = 0.05$? What is the P-value for this test?

(b) Comment on the practical significance versus the statistical significance of your answer.

10-61. The data shown below gave X = the water content of snow on April 1 and Y = the yield from April to July (in inches) on the Snake River watershed in Wyoming for 1919 to 1935. (The data were taken from an article in *Research Notes,* Vol. 61, 1950, Pacific Northwest Forest Range Experiment Station, Oregon)

x	y	x	y
23.1	10.5	37.9	22.8
32.8	16.7	30.5	14.1
31.8	18.2	25.1	12.9
32.0	17.0	12.4	8.8
30.4	16.3	35.1	17.4
24.0	10.5	31.5	14.9
39.5	23.1	21.1	10.5
24.2	12.4	27.6	16.1
52.5	24.9		

(a) Estimate the correlation between Y and X.

(b) Test the hypothesis that $\rho = 0$, using $\alpha = 0.05$.

(c) Fit a simple linear regression model and test for significance of regression using $\alpha = 0.05$. What conclusions can you draw? How is the test for significance of regression related to the test on ρ in part (b) above?

(d) Test the hypothesis H_0: $\beta_0 = 0$ versus H_1: $\beta_0 \neq 0$ and draw conclusions. Use $\alpha = 0.05$.

(e) Analyze the residuals and comment on model adequacy.

10-62. A random sample of $n = 25$ observations was made on the time to failure of an electronic component and the temperature in the application environment in which the component was used.

(a) Given that $r = 0.83$, test the hypothesis that $\rho = 0$, using $\alpha = 0.05$. What is the P-value for this test?

(b) Find a 95% confidence interval on ρ.

(c) Test the hypothesis H_0: $\rho = 0.8$ versus H_1: $\rho \neq 0.8$, using $\alpha = 0.05$. Find the P-value for this test.

Supplemental Exercises

10-63. Show that, for the simple linear regression model, the following statements are true:

(a) $\sum_{i=1}^{n} (y_i - \hat{y}_i) = 0$

(b) $\sum_{i=1}^{n} (y_i - \hat{y}_i)x_i = 0$

(c) $\frac{1}{n}\sum_{i=1}^{n} \hat{y}_i = \bar{y}$

10-64. An article in the *IEEE Transactions on Instrumentation and Measurement* ("Direct, Fast, and Accurate Measurement of V_T and K of MOS Transistor Using V_T-Sift Circuit," Vol. 40, 1991, pp. 951–955) described the use of a simple linear regression model to express drain current y (in mA) as a function of ground-to-source voltage x (in V). The data are as follows:

y	x
0.734	1.1
0.886	1.2
1.04	1.3
1.19	1.4
1.35	1.5
1.50	1.6
1.66	1.7
1.81	1.8
1.97	1.9
2.12	2.0

(a) Draw a scatter diagram of these data. Does a straight-line relationship seem plausible?

(b) Fit a simple linear regression model to these data.

(c) Test for significance of regression using $\alpha = 0.05$. What is the P-value for this test?

(d) Find a 95% confidence interval estimate on the slope.

(e) Test the hypothesis $H_0: \beta_0 = 0$ versus $H_1: \beta_0 \neq 0$ using $\alpha = 0.05$. What conclusions can you draw?

10-65. The strength of paper used in the manufacture of cardboard boxes (y) is related to the percentage of hardwood concentration in the original pulp (x). Under controlled conditions, a pilot plant manufactures 16 samples, each from a different batch of pulp, and measures the tensile strength. The data are shown here:

y	101.4	117.4	117.1	106.2
x	1.0	1.5	1.5	1.5
y	131.9	146.9	146.8	133.9
x	2.0	2.0	2.2	2.4
y	111.0	123.0	125.1	145.2
x	2.5	2.5	2.8	2.8
y	134.3	144.5	143.7	146.9
x	3.0	3.0	3.2	3.3

(a) Fit a simple linear regression model to the data.

(b) Test for significance of regression using $\alpha = 0.05$.

(c) Construct a 90% confidence interval on the slope β_1.

(d) Construct a 90% confidence interval on the intercept β_0.

(e) Construct a 95% confidence interval on the mean strength at $x = 2.5$.

(f) Test for lack of fit using $\alpha = 0.05$. What is the P-value for this test?

10-66. The following data are DC output from a windmill (y) and wind velocity (x).

Observation Number, i	Wind Velocity (MPH), x_i	DC Output y_i
1	5.00	1.582
2	6.00	1.822
3	3.40	1.057
4	2.70	0.500
5	10.00	2.236
6	9.70	2.386
7	9.55	2.294
8	3.05	0.558
9	8.15	2.166
10	6.20	1.866
11	2.90	0.653
12	6.35	1.930
13	4.60	1.562
14	5.80	1.737
15	7.40	2.088
16	3.60	1.137
17	7.85	2.179
18	8.80	2.112
19	7.00	1.800
20	5.45	1.501
21	9.10	2.303
22	10.20	2.310
23	4.10	1.194
24	3.95	1.144
25	2.45	0.123

(a) Draw a scatter diagram of these data. What type of relationship seems appropriate in relating y to x?

(b) Fit a simple linear regression model to these data.

(c) Test for significance of regression using $\alpha = 0.05$. What conclusions can you draw?

(d) Plot the residuals from the simple linear regression model versus $\hat{y}_i$ and versus wind velocity x. What do you conclude about model adequacy?

(e) Draw a scatter diagram of y versus $1/x$. What type of relationship seems appropriate?

(f) Fit a simple linear regression

model relating y to $1/x$. Test for significance of regression (use $\alpha = 0.05$), and plot the residuals from this model versus $\hat{y}_i$ and $1/x_i$. What can you conclude about model adequacy?

10-67. An electric utility is interested in developing a model relating peak hour demand (y in kw) to total monthly energy usage during the month (x, in kwh). Data for 50 residential customers are shown in the table in the right column.

(a) Draw a scatter diagram of y versus x.

(b) Fit the simple linear regression model.

(c) Test for significance of regression using $\alpha = 0.05$.

(d) Plot the residuals versus $\hat{y}_i$ and comment on the underlying regression assumptions. Specifically, does it seem that the equality of variance assumption is satisfied?

(e) Find a simple linear regression model using $\sqrt{y}$ as the response. Does this transformation on y stabilize the inequality of variance problem noted in part (d) above?

Customer	x(kwh)	y(kw)
1	679	0.79
2	292	0.44
3	1012	0.56
4	493	0.79
5	582	2.70
6	1156	3.64
7	997	4.73
8	2189	9.50
9	1097	5.34
10	2078	6.85
11	1818	5.84
12	1700	5.21
13	747	3.25
14	2030	4.43
15	1643	3.16
16	414	0.50
17	354	0.17
18	1276	1.88
19	745	0.77
20	795	3.70
21	540	0.56
22	874	1.56
23	1543	5.28
24	1029	0.64
25	710	4.00
26	1434	0.31
27	837	4.20
28	1748	4.88
29	1381	3.48
30	1428	7.58
31	1255	2.63
32	1777	4.99
33	370	0.59
34	2316	8.19
35	1130	4.79
36	463	0.51
37	770	1.74
38	724	4.10
39	808	3.94
40	790	0.96
41	783	3.29
42	406	0.44
43	1242	3.24
44	658	2.14
45	1746	5.71
46	895	4.12
47	1114	1.90
48	413	0.51
49	1787	8.33
50	3560	14.94

10-68. Consider the data shown below. Suppose that the relationship between Y and x is hypothesized to be $Y = (\beta_0 + \beta_1 x + \epsilon)^{-1}$. Fit an appropriate model to the data. Does the assumed model form seem reasonable?

x	10	15	18	12
y	0.1	0.13	0.09	0.15

x	9	8	11	6
y	0.20	0.21	0.18	0.24

10-69. Consider the weight and blood pressure data in Exercise 10-56. Fit a no-intercept model to the data, and compare it to the model obtained in Exercise 10-56. Which model is superior?

10-70. The data shown below, adapted from Montgomery and Peck (1992), present the number of certified mental defectives per 10,000 of estimated population in the United Kingdom (y) and the number of radio receiver licenses issued (x) by the BBC (in millions) for the years 1924 through 1937. Fit a regression model relating y and x. Comment on the model. Specifically, does the existence of a strong correlation imply a cause-and-effect relationship?

Year	Number of Certified Mental Defectives per 10,000 of Estimated U.K. Population (y)	Number of Radio Receiver Licenses Issued (millions) in the U.K. (x)
1924	8	1.350
1925	8	1.960
1926	9	2.270
1927	10	2.483
1928	11	2.730
1929	11	3.091
1930	12	3.674
1931	16	4.620
1932	18	5.497
1933	19	6.260
1934	20	7.012
1935	21	7.618
1936	22	8.131
1937	23	8.593

MIND-EXPANDING EXERCISES

10-71. Consider the simple linear regression model $Y = \beta_0 + \beta_1 x + \epsilon$, with $E(\epsilon) = 0$, $V(\epsilon) = \sigma^2$, and the errors ϵ uncorrelated.

(a) Show that $\text{cov}(\hat{\beta}_0, \hat{\beta}_1) = -\bar{x}\sigma^2/S_{xx}$.

(b) Show that $\text{cov}(\bar{Y}, \hat{\beta}_1) = 0$.

10-72. Consider the simple linear regression model $Y = \beta_0 + \beta_1 x + \epsilon$, with $E(\epsilon) = 0$, $V(\epsilon) = \sigma^2$, and the errors ϵ uncorrelated.

(a) Show that $E(\hat{\sigma}^2) = E(MS_E) = \sigma^2$.

(b) Show that $E(MS_R) = \sigma^2 + \beta_1^2 S_{xx}$.

10-73. Suppose that we have assumed the straight-line regression model

$$Y = \beta_0 + \beta_1 x_1 + \epsilon$$

but the response is affected by a second variable x_2 such that the true regression function is

$$E(Y) = \beta_0 + \beta_1 x_1 + \beta_2 x_2$$

Is the estimator of the slope in the simple linear regression model unbiased?

10-74. Suppose that we are fitting a line and we wish to make the variance of the regression coefficient $\hat{\beta}_1$ as small as possible. Where should the observations x_i, $i = 1, 2, \ldots, n$, be taken so as to minimize $V(\hat{\beta}_1)$? Discuss the practical implications of this allocation of the x_i.

10-75. **Weighted Least Squares.** Suppose that we are fitting the line $Y = \beta_0 +$

MIND-EXPANDING EXERCISES

$\beta_1 x + \epsilon$, but the variance of Y depends on the level of x; that is,

$$V(Y_i \mid x_i) = \sigma_i^2 = \frac{\sigma^2}{w_i} \quad i = 1, 2, \ldots, n$$

where the w_i are constants, often called *weights*. Show that for an appropriate objective function, the resulting least squares normal equations are

$$\hat{\beta}_0 \sum_{i=1}^{n} w_i + \hat{\beta}_1 \sum_{i=1}^{n} w_i x_i = \sum_{i=1}^{n} w_i y_i$$

$$\hat{\beta}_0 \sum_{i=1}^{n} w_i x_i + \hat{\beta}_1 \sum_{i=1}^{n} w_i x_i^2 = \sum_{i=1}^{n} w_i x_i y_i$$

10-76. Consider a situation where both Y and X are random variables. Let s_x and s_y be the sample standard deviations of the observed x's and y's, respectively. Show that an alternative expression for the fitted simple linear regression model $\hat{y} = \hat{\beta}_0 + \hat{\beta}_1 x$ is

$$\hat{y} = \bar{y} + r\frac{s_y}{s_x}(x - \bar{x})$$

10-77. Suppose that we are interested in fitting a simple linear regression model $Y = \beta_0 + \beta_1 x + \epsilon$, where the intercept, β_0, is known.

(a) Find the least squares estimator of β_1.

(b) What is the variance of the estimator of the slope in part (a)?

(c) Find an expression for a $100(1 - \alpha)$ percent confidence interval for the slope β_1. Is this interval longer than the corresponding interval for the case where both the intercept and slope are unknown? Justify your answer.

11 Multiple Linear Regression

CHAPTER OUTLINE

11-1 MULTIPLE LINEAR REGRESSION MODEL

Many applications of regression analysis involve situations in which there are more than one regressor variable. A regression model that contains more than one regressor variable is called a **multiple regression model.**

As an example, suppose that the effective life of a cutting tool depends on the cutting speed and the tool angle. A multiple regression model that might describe this relationship is

$$Y = \beta_0 + \beta_1 x_1 + \beta_2 x_2 + \epsilon \tag{11-1}$$

where Y represents the tool life, x_1 represents the cutting speed, x_2 represents the tool angle, and ϵ is a random error term. This is a **multiple linear regression model** with two regressors. The term *linear* is used because equation 11-1 is a linear function of the unknown parameters β_0, β_1, and β_2.

The regression model in equation 11-1 describes a plane in the three-dimensional space of Y, x_1, and x_2. Figure 11-1(a) shows this plane for the regression model

$$E(Y) = 50 + 10x_1 + 7x_2$$

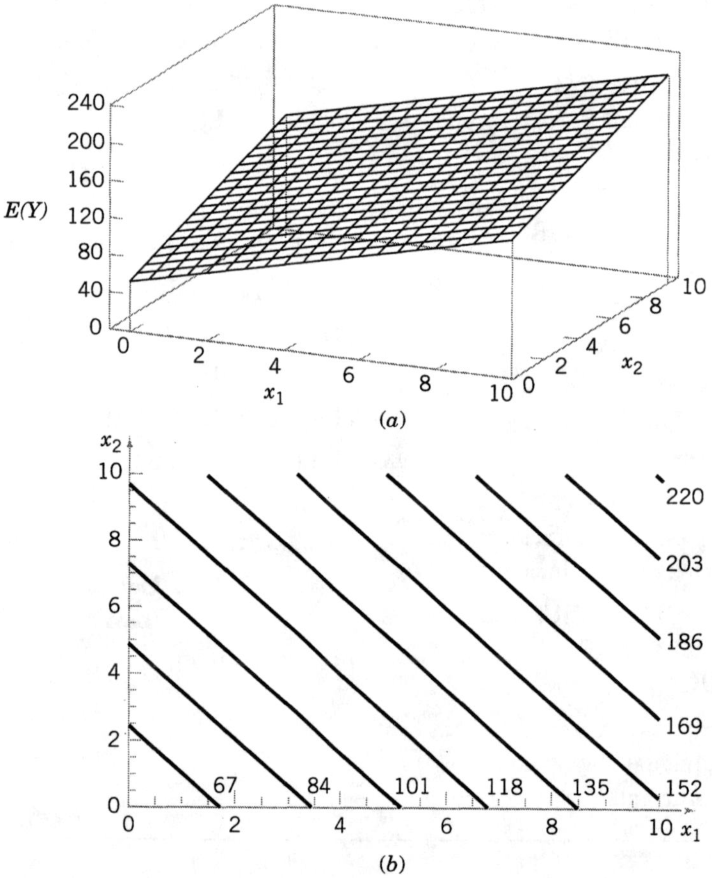

Figure 11-1 (a) The regression plane for the model $E(Y) = 50 + 10x_1 + 7x_2$. (b) The contour plot.

where we have assumed that the expected value of the error term is zero; that is $E(\epsilon) = 0$. The parameter β_0 is the **intercept** of the plane. We sometimes call β_1 and β_2 **partial regression coefficients,** because β_1 measures the expected change in Y per unit change in x_1 when x_2 is held constant, and β_2 measures the expected change in Y per unit change in x_2 when x_1 is held constant. Figure 11-1(b) shows a **contour plot** of the regression model—that is, lines of constant $E(Y)$ as a function of x_1 and x_2. Notice that the contour lines in this plot are straight lines.

In general, the **dependent variable** or **response** y may be related to k **independent** or **regressor variables.** The model

$$Y = \beta_0 + \beta_1 x_1 + \beta_2 x_2 + \cdots + \beta_k x_k + \epsilon \tag{11-2}$$

is called a multiple linear regression model with k regressor variables. The parameters $\beta_j, j = 0, 1, \ldots, k$, are called the regression coefficients. This model describes a hyperplane in the k-dimensional space of the regressor variables $\{x_j\}$. The parameter β_j represents the expected change in response Y per unit change in x_j when all the remaining regressors x_i $(i \neq j)$ are held constant.

Multiple linear regression models are often used as approximating functions. That is, the true functional relationship between Y and $x_1, x_2, \ldots, x_k$ is unknown, but over certain ranges of the independent variables the linear regression model is an adequate approximation.

Models that are more complex in structure than equation 11-2 may often still be analyzed by multiple linear regression techniques. For example, consider the cubic polynomial model in one regressor variable.

$$Y = \beta_0 + \beta_1 x + \beta_2 x^2 + \beta_3 x^3 + \epsilon \tag{11-3}$$

If we let $x_1 = x$, $x_2 = x^2$, $x_3 = x^3$, then equation 11-3 can be written as

$$Y = \beta_0 + \beta_1 x_1 + \beta_2 x_2 + \beta_3 x_3 + \epsilon \tag{11-4}$$

which is a multiple linear regression model with three regressor variables.

Models that include **interaction** effects may also be analyzed by multiple linear regression methods. An interaction between two variables can be represented by a cross-product term in the model, such as

$$Y = \beta_0 + \beta_1 x_1 + \beta_2 x_2 + \beta_{12} x_1 x_2 + \epsilon \tag{11-5}$$

If we let $x_3 = x_1 x_2$ and $\beta_3 = \beta_{12}$, then equation 11-5 can be written as

$$Y = \beta_0 + \beta_1 x_1 + \beta_2 x_2 + \beta_3 x_3 + \epsilon$$

which is a linear regression model.

Figure 11-2(a) and (b) shows the three-dimensional plot of the regression model

$$Y = 50 + 10x_1 + 7x_2 + 5x_1 x_2$$

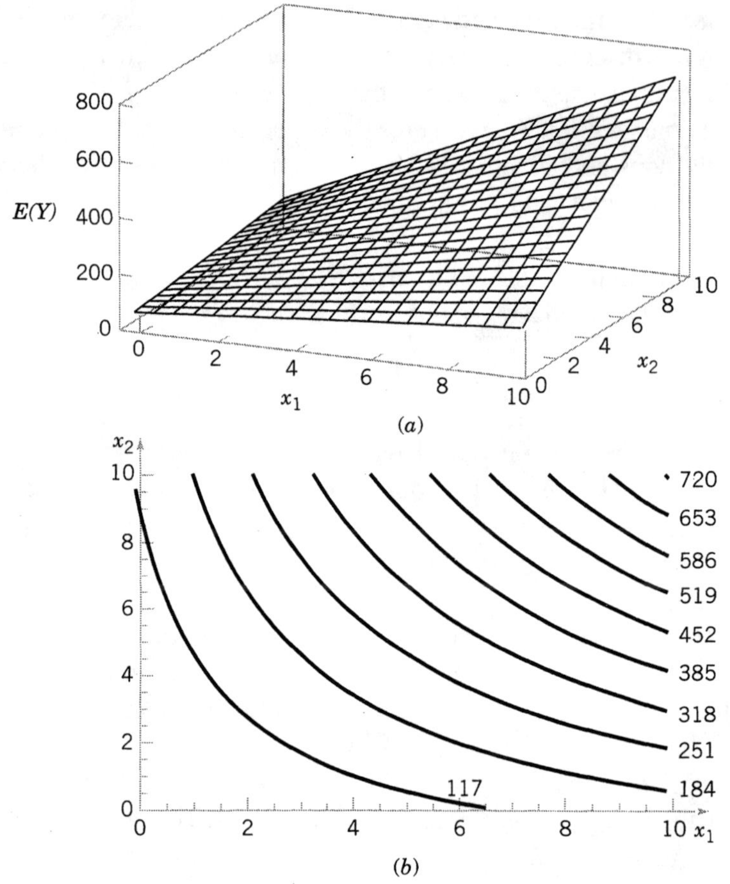

Figure 11-2 (a) Three-dimensional plot of regression model $E(Y) = 50 + 10x_1 + 7x_2 + 5x_1x_2$. (b) The contour plot.

and the corresponding two-dimensional contour plot. Notice that, although this model is a linear regression model, the shape of the surface that is generated by the model is not linear. In general, **any regression model that is linear in parameters** (the β's) **is a linear regression model, regardless of the shape of the surface that it generates.**

Figure 11-2 provides a nice graphical interpretation of an interaction. Generally, interaction implies that the effect produced by changing one variable (x_1, say) depends on the level of the other variable (x_2). For example, Fig. 11-2 shows that changing x_1 from 2 to 8 produces a much smaller change in $E(Y)$ when $x_2 = 2$ than when $x_2 = 10$. Interaction effects occur frequently in the study and analysis of real-world systems, and regression methods are one of the techniques that we can use to describe them.

As a final example, consider the second-order model with interaction

$$Y = \beta_0 + \beta_1 x_1 + \beta_2 x_2 + \beta_{11} x_1^2 + \beta_{22} x_2^2 + \beta_{12} x_1 x_2 + \epsilon \qquad (11\text{-}6)$$

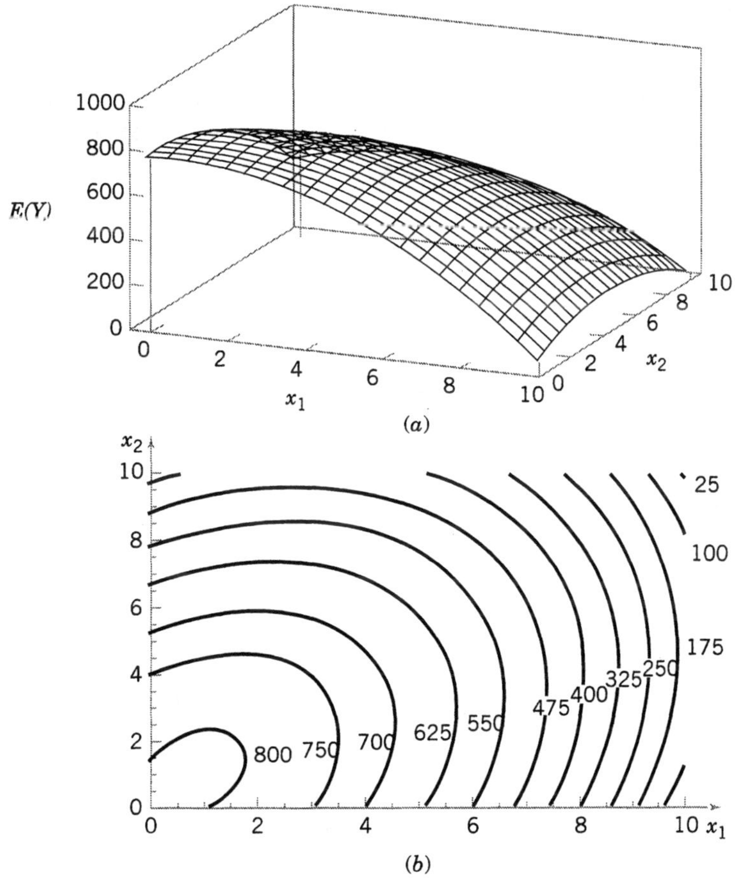

Figure 11-3 (a) Three-dimensional plot of the regression model $E(Y) = 800 + 10x_1 + 7x_2 - 8.5x_1^2 - 5x_2^2 + 4x_1x_2$. (b) The contour plot.

If we let $x_3 = x_1^2$, $x_4 = x_2^2$, $x_5 = x_1x_2$, $\beta_3 = \beta_{11}$, $\beta_4 = \beta_{22}$, and $\beta_5 = \beta_{12}$, then equation 11-6 can be written as a multiple linear regression model as follows:

$$Y = \beta_0 + \beta_1x_1 + \beta_2x_2 + \beta_3x_3 + \beta_4x_4 + \beta_5x_5 + \epsilon$$

Figure 11-3(a) and (b) show the three-dimensional plot and the corresponding contour plot for

$$E(Y) = 800 + 10x_1 + 7x_2 - 8.5x_1^2 - 5x_2^2 + 4x_1x_2$$

These plots indicate that the expected change in Y when x_1 is changed by one unit (say) is a function of *both* x_1 and x_2. The quadratic and interaction terms in this model produce a mound-shaped function. Depending on the values of the regression coefficients, the second-order model with interaction is capable of assuming a wide variety of shapes; thus, it is a very flexible regression model.

11-2 LEAST SQUARES ESTIMATION OF THE PARAMETERS

The method of least squares may be used to estimate the regression coefficients in the multiple regression model, equation 11-2. Suppose that $n > k$ observations are available, and let x_{ij} denote the ith observation or level of variable x_j. The observations are

$$(x_{i1}, x_{i2}, \ldots, x_{ik}, y_i), \qquad i = 1, 2, \ldots, n \quad \text{and} \quad n > k$$

It is customary to present the data for multiple regression in a table such as Table 11-1.

Each observation $(x_{i1}, x_{i2}, \ldots, x_{ik}, y_i)$, satisfies the model in equation 11-2, or

$$\begin{aligned} y_i &= \beta_0 + \beta_1 x_{i1} + \beta_2 x_{i2} + \cdots + \beta_k x_{ik} + \epsilon_i \\ &= \beta_0 + \sum_{j=1}^{k} \beta_j x_{ij} + \epsilon_i \qquad i = 1, 2, \ldots, n \end{aligned} \tag{11-7}$$

The least squares function is

$$L = \sum_{i=1}^{n} \epsilon_i^2 = \sum_{i=1}^{n} \left(y_i - \beta_0 - \sum_{j=1}^{k} \beta_j x_{ij} \right)^2 \tag{11-8}$$

The function L is to be minimized with respect to $\beta_0, \beta_1, \ldots, \beta_k$. The **least squares estimates** of $\beta_0, \beta_1, \ldots, \beta_k$ must satisfy

$$\left.\frac{\partial L}{\partial \beta_0}\right|_{\hat{\beta}_0, \hat{\beta}_1, \ldots, \hat{\beta}_k} = -2 \sum_{i=1}^{n} \left(y_i - \hat{\beta}_0 - \sum_{j=1}^{k} \hat{\beta}_j x_{ij} \right) = 0 \tag{11-9a}$$

and

$$\left.\frac{\partial L}{\partial \beta_j}\right|_{\hat{\beta}_0, \hat{\beta}_1, \ldots, \hat{\beta}_k} = -2 \sum_{i=1}^{n} \left(y_i - \hat{\beta}_0 - \sum_{j=1}^{k} \hat{\beta}_j x_{ij} \right) x_{ij} = 0 \qquad j = 1, 2, \ldots, k \tag{11-9b}$$

Table 11-1 Data for Multiple Linear Regression

y	x_1	x_2	$\ldots$	x_k
y_1	x_{11}	x_{12}	$\ldots$	x_{1k}
y_2	x_{21}	x_{22}	$\ldots$	x_{2k}
$\vdots$	$\vdots$	$\vdots$		$\vdots$
y_n	x_{n1}	x_{n2}	$\ldots$	x_{nk}

Simplifying equation 11-9, we obtain the **least squares normal equations**

$$
\begin{aligned}
n\hat{\beta}_0 + \hat{\beta}_1 \sum_{i=1}^{n} x_{i1} + \hat{\beta}_2 \sum_{i=1}^{n} x_{i2} + \cdots + \hat{\beta}_k \sum_{i=1}^{n} x_{ik} &= \sum_{i=1}^{n} y_i \\
\hat{\beta}_0 \sum_{i=1}^{n} x_{i1} + \hat{\beta}_1 \sum_{i=1}^{n} x_{i1}^2 + \hat{\beta}_2 \sum_{i=1}^{n} x_{i1}x_{i2} + \cdots + \hat{\beta}_k \sum_{i=1}^{n} x_{i1}x_{ik} &= \sum_{i=1}^{n} x_{i1}y_i \\
\vdots \qquad\qquad \vdots \qquad\qquad \vdots \qquad\qquad \vdots \qquad\qquad \vdots \\
\hat{\beta}_0 \sum_{i=1}^{n} x_{ik} + \hat{\beta}_1 \sum_{i=1}^{n} x_{ik}x_{i1} + \hat{\beta}_2 \sum_{i=1}^{n} x_{ik}x_{i2} + \cdots + \hat{\beta}_k \sum_{i=1}^{n} x_{ik}^2 &= \sum_{i=1}^{n} x_{ik}y_i
\end{aligned}
\tag{11-10}
$$

Note that there are $p = k + 1$ normal equations, one for each of the unknown regression coefficients. The solution to the normal equations will be the least squares estimators of the regression coefficients, $\hat{\beta}_0, \hat{\beta}_1, \ldots, \hat{\beta}_k$. The normal equations can be solved by any method appropriate for solving a system of linear equations.

EXAMPLE 11-1

In Chapter 1, we used data on pull strength of a wire bond in a semiconductor manufacturing process, wire length, and die height to illustrate building an empirical model. We will use the same data, repeated for convenience in Table 11-2, and show the details of estimating the model parameters. A three-dimensional scatter plot of the data is presented in Fig. 1-8.

Specifically, we will fit the multiple linear regression model

$$Y = \beta_0 + \beta_1 x_1 + \beta_2 x_2 + \epsilon$$

where Y = pull strength, x_1 = wire length, and x_2 = die height. From the data in Table 11-2 we calculate

$$n = 25, \quad \sum_{i=1}^{25} y_i = 725.82$$

$$\sum_{i=1}^{25} x_{i1} = 206, \quad \sum_{i=1}^{25} x_{i2} = 8{,}294$$

$$\sum_{i=1}^{25} x_{i1}^2 = 2{,}396, \quad \sum_{i=1}^{25} x_{i2}^2 = 3{,}531{,}848$$

$$\sum_{i=1}^{25} x_{i1}x_{i2} = 77{,}177, \quad \sum_{i=1}^{25} x_{i1}y_i = 8{,}008.37, \quad \sum_{i=1}^{25} x_{i2}y_i = 274{,}811.31$$

Table 11-2 Wire Bond Data for Example 11-1

Observation Number	Pull Strength y	Wire Length x_1	Die Height x_2
1	9.95	2	50
2	24.45	8	110
3	31.75	11	120
4	35.00	10	550
5	25.02	8	295
6	16.86	4	200
7	14.38	2	375
8	9.60	2	52
9	24.35	9	100
10	27.50	8	300
11	17.08	4	412
12	37.00	11	400
13	41.95	12	500
14	11.66	2	360
15	21.65	4	205
16	17.89	4	400
17	69.00	20	600
18	10.30	1	585
19	34.93	10	540
20	46.59	15	250
21	44.88	15	290
22	54.12	16	510
23	56.23	17	590
24	22.13	6	100
25	21.15	5	400

For the model $Y = \beta_0 + \beta_1 x_1 + \beta_2 x_2 + \epsilon$, the normal equations 11-10 are

$$n\hat{\beta}_0 + \hat{\beta}_1 \sum_{i=1}^{n} x_{i1} + \hat{\beta}_2 \sum_{i=1}^{n} x_{i2} = \sum_{i=1}^{n} y_i$$

$$\hat{\beta}_0 \sum_{i=1}^{n} x_{i1} + \hat{\beta}_1 \sum_{i=1}^{n} x_{i1}^2 + \hat{\beta}_2 \sum_{i=1}^{n} x_{i1}x_{i2} = \sum_{i=1}^{n} x_{i1} y_i$$

$$\hat{\beta}_0 \sum_{i=1}^{n} x_{i2} + \hat{\beta}_1 \sum_{i=1}^{n} x_{i1}x_{i2} + \hat{\beta}_2 \sum_{i=1}^{n} x_{i2}^2 = \sum_{i=1}^{n} x_{i2} y_i$$

Inserting the computed summations into the normal equations, we obtain

$$\begin{aligned} 25\hat{\beta}_0 + \quad 206\hat{\beta}_1 + \quad 8294\hat{\beta}_2 &= 725.82 \\ 206\hat{\beta}_0 + \quad 2396\hat{\beta}_1 + \quad 77{,}177\hat{\beta}_2 &= 8{,}008.37 \\ 8294\hat{\beta}_0 + 77{,}177\hat{\beta}_1 + 3{,}531{,}848\hat{\beta}_2 &= 274{,}811.31 \end{aligned}$$

The solution of this set of equations is

$$\hat{\beta}_0 = 2.26379, \qquad \hat{\beta}_1 = 2.74427, \qquad \hat{\beta}_2 = 0.01253$$

Therefore, the fitted regression equation is

$$\hat{y} = 2.26379 + 2.74427x_1 + 0.01253x_2$$

This equation can be used to predict pull strength for pairs of values of the regressor variables wire length (x_1) and die height (x_2). This is essentially the same regression model given in equation 1-12, Section 1-3. Figure 1-9 shows a plot of the predicted values $\hat{y}$ generated from this equation.

11-3 MATRIX APPROACH TO MULTIPLE LINEAR REGRESSION

In fitting a multiple regression model, it is much more convenient to express the mathematical operations using matrix notation. Suppose that there are k regressor variables and n observations, $(x_{i1}, x_{i2}, \ldots, x_{ik}, y_i)$, $i = 1 = 1, 2, \ldots, n$ and that the model relating the regressors to the response is

$$y_i = \beta_0 + \beta_1 x_{i1} + \beta_2 x_{i2} + \cdots + \beta_k x_{ik} + \epsilon_i \qquad i = 1, 2, \ldots, n$$

This model is a system of n equations that can be expressed in matrix notation as

$$\mathbf{y} = \mathbf{X}\boldsymbol{\beta} + \boldsymbol{\epsilon} \tag{11-11}$$

where

$$\mathbf{y} = \begin{bmatrix} y_1 \\ y_2 \\ \vdots \\ y_n \end{bmatrix} \qquad \mathbf{X} = \begin{bmatrix} 1 & x_{11} & x_{12} & \cdots & x_{1k} \\ 1 & x_{21} & x_{22} & \cdots & x_{2k} \\ \vdots & \vdots & \vdots & & \vdots \\ 1 & x_{n1} & x_{n2} & \cdots & x_{nk} \end{bmatrix}$$

$$\boldsymbol{\beta} = \begin{bmatrix} \beta_0 \\ \beta_1 \\ \vdots \\ \beta_k \end{bmatrix} \quad \text{and} \quad \boldsymbol{\epsilon} = \begin{bmatrix} \epsilon_1 \\ \epsilon_2 \\ \vdots \\ \epsilon_n \end{bmatrix}$$

In general, $\mathbf{y}$ is an $(n \times 1)$ vector of the observations, $\mathbf{X}$ is an $(n \times p)$ matrix of the levels of the independent variables, $\boldsymbol{\beta}$ is a $(p \times 1)$ vector of the regression coefficients, and $\boldsymbol{\epsilon}$ is a $(n \times 1)$ vector of random errors.

We wish to find the vector of least squares estimators, $\hat{\boldsymbol{\beta}}$, that minimizes

$$L = \sum_{i=1}^{n} \epsilon_i^2 = \boldsymbol{\epsilon}'\boldsymbol{\epsilon} = (\mathbf{y} - \mathbf{X}\boldsymbol{\beta})'(\mathbf{y} - \mathbf{X}\boldsymbol{\beta})$$

The least squares estimator $\hat{\boldsymbol{\beta}}$ is the solution for $\boldsymbol{\beta}$ in the equations

$$\frac{\partial L}{\partial \boldsymbol{\beta}} = \mathbf{0}$$

We will not give the details of taking the derivatives above; however, the resulting equations that must be solved are

$$\mathbf{X}'\mathbf{X}\hat{\boldsymbol{\beta}} = \mathbf{X}'\mathbf{y} \tag{11-12}$$

Equations 11-12 are the least squares normal equations in matrix form. They are identical to the scalar form of the normal equations given earlier in equations 11-10. To solve the normal equations, multiply both sides of equations 11-12 by the inverse of $\mathbf{X}'\mathbf{X}$. Therefore, the least squares estimate of $\boldsymbol{\beta}$ is

$$\hat{\boldsymbol{\beta}} = (\mathbf{X}'\mathbf{X})^{-1}\mathbf{X}'\mathbf{y} \tag{11-13}$$

Note that there are $p = k + 1$ normal equations in $p = k + 1$ unknowns (the values of $\hat{\beta}_0, \hat{\beta}_1, \ldots, \hat{\beta}_k$). Furthermore, the matrix $\mathbf{X}'\mathbf{X}$ is often nonsingular, as was assumed above, so that the methods described in textbooks on determinants and matrices for inverting these matrices can be used to find $(\mathbf{X}'\mathbf{X})^{-1}$. In practice, multiple regression calculations are almost always performed using a computer.

It is easy to see that the matrix form of the normal equations is identical to the scalar form. Writing out equation 11-12 in detail, we obtain

$$\begin{bmatrix} n & \sum_{i=1}^{n} x_{i1} & \sum_{i=1}^{n} x_{i2} & \cdots & \sum_{i=1}^{n} x_{ik} \\ \sum_{i=1}^{n} x_{i1} & \sum_{i=1}^{n} x_{i1}^2 & \sum_{i=1}^{n} x_{i1}x_{i2} & \cdots & \sum_{i=1}^{n} x_{i1}x_{ik} \\ \vdots & \vdots & \vdots & & \vdots \\ \sum_{i=1}^{n} x_{ik} & \sum_{i=1}^{n} x_{ik}x_{i1} & \sum_{i=1}^{n} x_{ik}x_{i2} & \cdots & \sum_{i=1}^{n} x_{ik}^2 \end{bmatrix} \begin{bmatrix} \hat{\beta}_0 \\ \hat{\beta}_1 \\ \vdots \\ \hat{\beta}_k \end{bmatrix} = \begin{bmatrix} \sum_{i=1}^{n} y_i \\ \sum_{i=1}^{n} x_{i1} y_i \\ \vdots \\ \sum_{i=1}^{n} x_{ik} y_i \end{bmatrix}$$

If the indicated matrix multiplication is performed, the scalar form of the normal equations (that is, equation 11-10) will result. In this form it is easy to see that $\mathbf{X}'\mathbf{X}$ is a $(p \times p)$ symmetric matrix and $\mathbf{X}'\mathbf{y}$ is a $(p \times 1)$ column vector. Note the special structure of the $\mathbf{X}'\mathbf{X}$ matrix. The diagonal elements of $\mathbf{X}'\mathbf{X}$ are the sums of squares of the elements in the columns of **X,** and the off-diagonal elements are the sums of cross-products of the elements in the columns of **X.** Furthermore, note that the elements of $\mathbf{X}'\mathbf{y}$ are the sums of cross-products of the columns of **X** and the observations $\{y_i\}$.

The fitted regression model is

$$\hat{y}_i = \hat{\beta}_0 + \sum_{j=1}^{k} \hat{\beta}_j x_{ij} \qquad i = 1, 2, \ldots, n \tag{11-14}$$

In matrix notation, the fitted model is

$$\hat{\mathbf{y}} = \mathbf{X}\hat{\boldsymbol{\beta}}$$

The difference between the observation y_i and the fitted value $\hat{y}_i$ is a **residual,** say, $e_i = y_i - \hat{y}_i$. The $(n \times 1)$ vector of residuals is denoted by

$$\mathbf{e} = \mathbf{y} - \hat{\mathbf{y}} \tag{11-15}$$

EXAMPLE 11-2

In Example 11-1, we illustrated fitting the multiple regression model

$$y = \beta_0 + \beta_1 x_1 + \beta_2 x_2 + \epsilon$$

where y is the observed pull strength for a wire bond, x_1 is the wire length, and x_2 is the die height. The 25 observations are in Table 11-2. We will now use the matrix approach to fit the regression model above to these data. The **X** matrix and **y** vector for this model are

$$
\mathbf{X} = \begin{bmatrix} 1 & 2 & 50 \\ 1 & 8 & 110 \\ 1 & 11 & 120 \\ 1 & 10 & 550 \\ 1 & 8 & 295 \\ 1 & 4 & 200 \\ 1 & 2 & 375 \\ 1 & 2 & 52 \\ 1 & 9 & 100 \\ 1 & 8 & 300 \\ 1 & 4 & 412 \\ 1 & 11 & 400 \\ 1 & 12 & 500 \\ 1 & 2 & 360 \\ 1 & 4 & 205 \\ 1 & 4 & 400 \\ 1 & 20 & 600 \\ 1 & 1 & 585 \\ 1 & 10 & 540 \\ 1 & 15 & 250 \\ 1 & 15 & 290 \\ 1 & 16 & 510 \\ 1 & 17 & 590 \\ 1 & 6 & 100 \\ 1 & 5 & 400 \end{bmatrix} \qquad \mathbf{y} = \begin{bmatrix} 9.95 \\ 24.45 \\ 31.75 \\ 35.00 \\ 25.02 \\ 16.86 \\ 14.38 \\ 9.60 \\ 24.35 \\ 27.50 \\ 17.08 \\ 37.00 \\ 41.95 \\ 11.66 \\ 21.65 \\ 17.89 \\ 69.00 \\ 10.30 \\ 34.93 \\ 46.59 \\ 44.88 \\ 54.12 \\ 56.63 \\ 22.13 \\ 21.15 \end{bmatrix}
$$

The $\mathbf{X}'\mathbf{X}$ matrix is

$$
\mathbf{X}'\mathbf{X} = \begin{bmatrix} 1 & 1 & \cdots & 1 \\ 2 & 8 & \cdots & 5 \\ 50 & 110 & \cdots & 400 \end{bmatrix} \begin{bmatrix} 1 & 2 & 50 \\ 1 & 8 & 110 \\ \vdots & \vdots & \vdots \\ 1 & 5 & 400 \end{bmatrix}
$$

$$
\begin{bmatrix} 25 & 206 & 8{,}294 \\ 206 & 2{,}396 & 77{,}177 \\ 8{,}294 & 77{,}177 & 3{,}531{,}848 \end{bmatrix}
$$

and the $\mathbf{X}'\mathbf{y}$ vector is

$$
\mathbf{X}'\mathbf{y} = \begin{bmatrix} 1 & 1 & \cdots & 1 \\ 2 & 8 & \cdots & 5 \\ 50 & 110 & \cdots & 400 \end{bmatrix} \begin{bmatrix} 9.95 \\ 24.45 \\ \vdots \\ 21.15 \end{bmatrix} = \begin{bmatrix} 725.82 \\ 8{,}008.37 \\ 274{,}811.31 \end{bmatrix}
$$

The least squares estimates are found from equation 11-13 as

$$\hat{\boldsymbol{\beta}} = (\mathbf{X}'\mathbf{X})^{-1}\mathbf{X}'\mathbf{y}$$

or

$$\begin{bmatrix} \hat{\beta}_0 \\ \hat{\beta}_1 \\ \hat{\beta}_2 \end{bmatrix} = \begin{bmatrix} 25 & 206 & 8{,}294 \\ 206 & 2{,}396 & 77{,}177 \\ 8{,}294 & 77{,}177 & 3{,}531{,}848 \end{bmatrix}^{-1} \begin{bmatrix} 725.82 \\ 8{,}008.37 \\ 274{,}811.31 \end{bmatrix}$$

$$= \begin{bmatrix} 0.21653 & -0.007491 & -0.000340 \\ -0.007491 & 0.001671 & -0.000019 \\ -0.000340 & -0.000019 & -0.0000015 \end{bmatrix} \begin{bmatrix} 725.82 \\ 8{,}008.47 \\ 274{,}811.31 \end{bmatrix}$$

$$= \begin{bmatrix} 2.26379143 \\ 2.74426964 \\ 0.01252781 \end{bmatrix}$$

Therefore, the fitted regression model with the regression coefficients rounded to five decimal places is

$$\hat{y} = 2.26379 + 2.74427x_1 + 0.01253x_2$$

This is identical to the results obtained in Example 11-1.

This regression model can be used to predict values of pull strength for various values of wire length (x_1) and die height (x_2). We can also obtain the **fitted values** $\hat{y}_i$ by substituting each observation (x_{i1}, x_{i2}), $i = 1, 2, \ldots, n$, into the equation. For example, the first observation has $x_{11} = 2$ and $x_{12} = 50$, and the fitted value is

$$\begin{aligned} \hat{y}_1 &= 2.26379 + 2.74427x_{11} + 0.01253x_{12} \\ &= 2.26379 + 2.74427(2) + 0.01253(50) \\ &= 8.38 \end{aligned}$$

The corresponding observed value is $y_1 = 9.95$. The *residual* corresponding to the first observation is

$$\begin{aligned} e_1 &= y_1 - \hat{y}_1 \\ &= 9.95 - 8.38 \\ &= 1.57 \end{aligned}$$

Table 11-3 displays all 25 fitted values $\hat{y}_i$ and the corresponding residuals. The fitted values and residuals are calculated to the same accuracy as the original data.

Table 11-3 Observations, Fitted Values, and Residuals for Example 11-2

Observation Number	y_i	$\hat{y}_i$	$e_i = y_i - \hat{y}_i$
1	9.95	8.38	1.57
2	24.45	25.60	−1.15
3	31.75	33.95	−2.20
4	35.00	36.60	−1.60
5	25.02	27.91	−2.89
6	16.86	15.75	1.11
7	14.38	12.45	1.93
8	9.60	8.40	1.20
9	24.35	28.21	−3.86
10	27.50	27.98	−0.48
11	17.08	18.40	−1.32
12	37.00	37.46	−0.46
13	41.95	41.46	0.49
14	11.66	12.26	−0.60
15	21.65	15.81	5.84
16	17.89	18.25	−0.36
17	69.00	64.67	4.33
18	10.30	12.34	−2.04
19	34.93	36.47	−1.54
20	46.59	46.56	−0.03
21	44.88	47.06	−2.18
22	54.12	52.56	1.56
23	56.63	56.31	0.32
24	22.13	19.98	2.15
25	21.15	21.00	0.15

Computers are almost always used in fitting multiple regression models. Table 11-4 presents some of the output from Minitab for the wire bond pull strength data in Examples 11-1 and 11-2. Notice that the upper part of the table contains the numerical estimates of the regression coefficients. The computer also calculates several other quantities that reflect important information about the regression model. In subsequent sections, we will define and explain the quantities in this output.

Table 11-4 Minitab Output for the Wire Bond Pull Strength Data

Regression Analysis

The regression equation is
$y = 2.26 + 2.74x_1 + 0.0125x_2$

Predictor	Coef	StDev	T	P
Constant	2.264	1.060	2.14	0.044
x_1	2.74427	0.09352	29.34	0.000
x_2	0.012528	0.002798	4.48	0.000

S = 2.288 R-Sq = 98.1% R-Sq(adj) = 97.9%

Analysis of Variance

Source	DF	SS	MS	F	P
Regression	2	5990.8	2995.4	572.17	0.000
Error	22	115.2	5.2		
Total	24	6105.9			

Source	DF	Seq SS
x_1	1	5885.9
x_2	1	104.9

EXERCISES FOR SECTIONS 11-2 AND 11-3

11-1. A study was performed to investigate the shear strength of soil (y) as it related to depth in feet (x_1) and moisture content (x_2). Ten observations were collected, and the following summary quantities obtained: $n = 10$, $\Sigma x_{i1} = 223$, $\Sigma x_{i2} = 553$, $\Sigma y_i = 1{,}916$, $\Sigma x_{i1}^2 = 5{,}200.9$, $\Sigma x_{i2}^2 = 31{,}729$, $\Sigma x_{i1}x_{i2} = 12{,}352$, $\Sigma x_{i1}y_i = 43{,}550.8$, $\Sigma x_{i2}y_i = 104{,}736.8$, and $\Sigma y_i^2 = 371{,}595.6$.

(a) Set up the least squares normal equations for the model $Y = \beta_0 + \beta_1 x_1 + \beta_2 x_2 + \epsilon$.

(b) Estimate the parameters in the model in part (a).

(c) What is the predicted strength when $x_1 = 18$ feet and $x_2 = 43\%$?

11-2. A regression model is to be developed for predicting the ability of soil to absorb chemical contaminants. Ten observations have been taken on a soil absorption index (y) and two regressors: x_1 = amount of extractable iron ore and x_2 = amount of bauxite. We wish to fit the model $Y = \beta_0 + \beta_1 x_1 + \beta_2 x_2 + \epsilon$. Some necessary quantities are:

$$(\mathbf{X}'\mathbf{X})^{-1} = \begin{bmatrix} 1.17991 & -7.30982\text{ E-3} & 7.3006\text{ E-4} \\ -7.30982\text{ E-3} & 7.9799\text{ E-5} & -1.23713\text{ E-4} \\ 7.3006\text{ E-4} & -1.23713\text{ E-4} & 4.6576\text{ E-4} \end{bmatrix}, \quad \mathbf{X}'\mathbf{y} = \begin{bmatrix} 220 \\ 36{,}768 \\ 9{,}965 \end{bmatrix}$$

(a) Estimate the regression coefficients in the model specified above.

(b) What is the predicted value of the absorption index y when $x_1 = 200$ and $x_2 = 50$?

11-3. A chemical engineer is investigating how the amount of conversion of a product from a raw material (y) depends on reaction temperature (x_1) and the reaction time (x_2). He has developed the following regression models:

1. $\hat{y} = 100 + 2x_1 + 4x_2$

2. $\hat{y} = 95 + 1.5x_1 + 3x_2 + 2x_1x_2$

Both models have been built over the range $0.5 \leq x_2 \leq 10$.

(a) What is the predicted value of conversion when $x_2 = 2$? Repeat this calculation for $x_2 = 8$. Draw a graph of the predicted values for both conversion models. Comment on the effect of the interaction term in model 2.

(b) Find the expected change in the mean conversion for a unit change in temperature x_1 for model 1 when $x_2 = 5$. Does this quantity depend on the specific value of reaction time selected? Why?

(c) Find the expected change in the mean conversion for a unit change in temperature x_1 for model 2 when $x_2 = 5$. Repeat this calculation for $x_2 = 2$ and $x_2 = 8$. Does the result depend on the value selected for x_2? Why?

11-4. The data in the table on page 499 are the 1976 team performance statistics for the teams in the National Football League (*Source: The Sporting News*).

(a) Fit a multiple regression model relating the number of games won to the teams' passing yardage (x_2), the percent rushing plays (x_7), and the opponents' yards rushing (x_8).

(b) Use the model to predict the number of games won when $x_2 = 2000$ yards, $x_7 = 60\%$, and $x_8 = 1800$ yards.

11-5. The table on page 500 presents gasoline mileage performance for 25 automobiles (*Source: Motor Trend,* 1975).

(a) Fit a multiple regression model relating gasoline mileage to engine displacement (x_1) and number of carburetor barrels (x_6).

(b) Use the model developed in part (a) to predict mileage performance when the engine has displacement $x_1 = 300$ in^3 and a four-barrel carburetor.

11-6. The electric power consumed each month by a chemical plant is thought to be related to the average ambient temperature (x_1), the number of days in the month (x_2), the average product purity (x_3), and the tons of product produced (x_4). The past year's historical data are available and are presented in the following table:

y	x_1	x_2	x_3	x_4
240	25	24	91	100
236	31	21	90	95
290	45	24	88	110
274	60	25	87	88
301	65	25	91	94
316	72	26	94	99
300	80	25	87	97
296	84	25	86	96
267	75	24	88	110
276	60	25	91	105
288	50	25	90	100
261	38	23	89	98

(a) Fit a multiple linear regression model to these data.

(b) Predict power consumption for a month in which $x_1 = 75$°F, $x_2 = 24$ days, $x_3 = 90\%$, and $x_4 = 98$ tons.

National Football League 1976 Team Performance

Team	y	x_1	x_2	x_3	x_4	x_5	x_6	x_7	x_8	x_9
Washington	10	2113	1985	38.9	64.7	+4	868	59.7	2205	1917
Minnesota	11	2003	2855	38.8	61.3	+3	615	55.0	2096	1575
New England	11	2957	1737	40.1	60.0	+14	914	65.6	1847	2175
Oakland	13	2285	2905	41.6	45.3	−4	957	61.4	1903	2476
Pittsburgh	10	2971	1666	39.2	53.8	+15	836	66.1	1457	1866
Baltimore	11	2309	2927	39.7	74.1	+8	786	61.0	1848	2339
Los Angeles	10	2528	2341	38.1	65.4	+12	754	66.1	1564	2092
Dallas	11	2147	2737	37.0	78.3	−1	797	58.9	2476	2254
Atlanta	4	1689	1414	42.1	47.6	−3	714	57.0	2577	2001
Buffalo	2	2566	1838	42.3	54.2	−1	797	58.9	2476	2254
Chicago	7	2363	1480	37.3	48.0	+19	984	68.5	1984	2217
Cincinnati	10	2109	2191	39.5	51.9	+6	819	59.2	1901	1686
Cleveland	9	2295	2229	37.4	53.6	−5	1037	58.8	1761	2032
Denver	9	1932	2204	35.1	71.4	+3	986	58.6	1709	2025
Detroit	6	2213	2140	38.8	58.3	+6	819	59.2	1901	1686
Green Bay	5	1722	1730	36.6	52.6	−19	791	54.4	2288	1835
Houston	5	1498	2072	35.3	59.3	−5	776	49.6	2072	1914
Kansas City	5	1873	2929	41.1	55.3	+10	789	54.3	2861	2496
Miami	6	2118	2268	38.6	69.6	+6	582	58.7	2411	2670
New Orleans	4	1775	1983	39.3	78.3	+7	901	51.7	2289	2202
New York Giants	3	1904	1792	39.7	38.1	−9	734	61.9	2203	1988
New York Jets	3	1929	1606	39.7	68.8	−21	627	52.7	2592	2324
Philadelphia	4	2080	1492	35.5	68.8	−8	722	57.8	2053	2550
St. Louis	10	2301	2835	35.3	74.1	+2	683	59.7	1979	2110
San Diego	6	2040	2416	38.7	50.0	0	576	54.9	2048	2628
San Francisco	8	2447	1638	39.9	57.1	−8	848	65.3	1786	1776
Seattle	2	1416	2649	37.4	56.3	−22	684	43.8	2876	2524
Tampa Bay	0	1503	1503	39.3	47.0	−9	875	53.5	2560	2241

y: Games won (per 14 game season)
x_1: Rushing yards (season)
x_2: Passing yards (season)
x_3: Punting yards (yds/punt)
x_4: Field goal percentage (Fgs made/Fgs attempted—season)
x_5: Turnover differential (turnovers acquired − turnovers lost)
x_6: Penalty yards (season)
x_7: Percent rushing (rushing plays/total plays)
x_8: Opponents' rushing yards (season)
x_9: Opponents' passing yards (season)

Gasoline Mileage Performance for 25 Automobiles

Automobile	y	x_1	x_2	x_3	x_4	x_5	x_6	x_7	x_8	x_9	x_{10}	x_{11}
Apollo	18.90	350	165	260	8.0:1	2.56:1	4	3	200.3	69.9	3910	A
Nova	20.00	250	105	185	8.25:1	2.73:1	1	3	196.7	72.2	3510	A
Monarch	18.25	351	143	255	8.0:1	3.00:1	2	3	199.9	74.0	3890	A
Duster	20.07	225	95	170	8.4:1	2.76:1	1	3	194.1	71.8	3365	M
Jenson Conv.	11.2	440	215	330	8.2:1	2.88:1	4	3	184.5	69	4215	A
Skyhawk	22.12	231	110	175	8.0:1	2.56:1	2	3	179.3	65.4	3020	A
Scirocco	34.70	89.7	70	81	8.2:1	3.90:1	2	4	155.7	64	1905	M
Corolla SR-5	30.40	96.9	75	83	9.0:1	4.30:1	2	5	165.2	65	2320	M
Camaro	16.50	350	155	250	8.5:1	3.08:1	4	3	195.4	74.4	3885	A
Datsun B210	36.50	85.3	80	83	8.5:1	3.89:1	2	4	160.6	62.2	2009	M
Capri II	21.50	171	109	146	8.2:1	3.22:1	2	4	170.4	66.9	2655	M
Pacer	19.70	258	110	195	8.0:1	3.08:1	1	3	171.5	77	3375	A
Granada	17.80	302	129	220	8.0:1	3.0:1	2	3	199.9	74	3890	A
Eldorado	14.39	500	190	360	8.5:1	2.73:1	4	3	224.1	79.8	5290	A
Imperial	14.89	440	215	330	8.2:1	2.71:1	4	3	231.0	79.7	5185	A
Nova LN	17.80	350	155	250	8.5:1	3.08:1	4	3	196.7	72.2	3910	A
Starfire	23.54	231	110	175	8.0:1	2.56:1	2	3	179.3	65.4	3050	A
Cordoba	21.47	360	180	290	8.4:1	2.45:1	2	3	214.2	76.3	4250	A
Trans Am	16.59	400	185	NA	7.6:1	3.08:1	4	3	196	73	3850	A
Corolla E-5	31.90	96.9	75	83	9.0:1	4.30:1	2	5	165.2	61.8	2275	M
Mark IV	13.27	460	223	366	8.0:1	3.00:1	4	3	228	79.8	5430	A
Celica GT	23.90	133.6	96	120	8.4:1	3.91:1	2	5	171.5	63.4	2535	M
Charger SE	19.73	318	140	255	8.5:1	2.71:1	2	3	215.3	76.3	4370	A
Cougar	13.90	351	148	243	8.0:1	3.25:1	2	3	215.5	78.5	4540	A
Corvette	16.50	350	165	255	8.5:1	2.73:1	4	3	185.2	69	3660	A

y: Miles/gallon
x_1: Displacement (cubic in.)
x_2: Horsepower (ft-lb)
x_3: Torque (ft-lb)
x_4: Compression ratio
x_5: Rear axle ratio
x_6: Carburetor (barrels)
x_7: No. of transmission speeds
x_8: Overall length (in.)
x_9: Width (in.)
x_{10}: Weight (lbs)
x_{11}: Type of transmission (A—automatic, M—manual)

11-7. A study was performed on wear of a bearing y and its relationship to x_1 = oil viscosity and x_2 = load. The following data were obtained.

y	x_1	x_2
193	1.6	851
230	15.5	816
172	22.0	1058
91	43.0	1201
113	33.0	1357
125	40.0	1115

(a) Fit a multiple linear regression model to these data.

(b) Use the model to predict wear when $x_1 = 25$ and $x_2 = 1000$.

(c) Fit a multiple linear regression model with an interaction term to these data.

(d) Use the model in (c) to predict when $x_1 = 25$ and $x_2 = 1000$. Compare this prediction with the predicted value from part (b) above.

11-8. The pull strength of a wire bond is an important characteristic. The table below gives information on pull strength (y), die height (x_1), post height (x_2), loop height (x_3), wire length (x_4), bond width on the die (x_5), and bond width on the post (x_6).

y	x_1	x_2	x_3	x_4	x_5	x_6
8.0	5.2	19.6	29.6	94.9	2.1	2.3
8.3	5.2	19.8	32.4	89.7	2.1	1.8
8.5	5.8	19.6	31.0	96.2	2.0	2.0
8.8	6.4	19.4	32.4	95.6	2.2	2.1
9.0	5.8	18.6	28.6	86.5	2.0	1.8
9.3	5.2	18.8	30.6	84.5	2.1	2.1
9.3	5.6	20.4	32.4	88.8	2.2	1.9
9.5	6.0	19.0	32.6	85.7	2.1	1.9
9.8	5.2	20.8	32.2	93.6	2.3	2.1
10.0	5.8	19.9	31.8	86.0	2.1	1.8
10.3	6.4	18.0	32.6	87.1	2.0	1.6
10.5	6.0	20.6	33.4	93.1	2.1	2.1
10.8	6.2	20.2	31.8	83.4	2.2	2.1
11.0	6.2	20.2	32.4	94.5	2.1	1.9
11.3	6.2	19.2	31.4	83.4	1.9	1.8
11.5	5.6	17.0	33.2	85.2	2.1	2.1
11.8	6.0	19.8	35.4	84.1	2.0	1.8
12.3	5.8	18.8	34.0	86.9	2.1	1.8
12.5	5.6	18.6	34.2	83.0	1.9	2.0

(a) Fit a multiple linear regression model using x_2, x_3, x_4, and x_5 as the regressors.

(b) Use the model from part (a) to predict pull strength when $x_2 = 20$, $x_3 = 30$, $x_4 = 90$, and $x_5 = 2.0$.

11-9. An engineer at a semiconductor company wants to model the relationship between the device HFE (y) and three parameters: Emitter-RS (x_1), Base-RS (x_2), and Emitter-to-Base RS (x_3). The data are shown in the following table.

x_1 Emitter-RS	x_2 Base-RS	x_3 B-E-RS	y HFE-1M-5V
14.620	226.00	7.000	128.40
15.630	220.00	3.375	52.62
14.620	217.40	6.375	113.90
15.000	220.00	6.000	98.01
14.500	226.50	7.625	139.90
15.250	224.10	6.000	102.60
16.120	220.50	3.375	48.14
15.130	223.50	6.125	109.60
15.500	217.60	5.000	82.68
15.130	228.50	6.625	112.60
15.500	230.20	5.750	97.52
16.120	226.50	3.750	59.06
15.130	226.60	6.125	111.80
15.630	225.60	5.375	89.09
15.380	229.70	5.875	101.00
14.380	234.00	8.875	171.90
15.500	230.00	4.000	66.80
14.250	224.30	8.000	157.10
14.500	240.50	10.870	208.40
14.620	223.70	7.375	133.40

(a) Fit a multiple linear regression model to the data.

(b) Predict HFE when $x_1 = 14.5$, $x_2 = 220$, and $x_3 = 5.0$.

11-10. Heat treating is often used to carburize metal parts, such as gears. The thickness of the carburized layer is considered a crucial feature of the gear and contributes to the overall reliability of the part. Because of the critical nature of this feature, two different lab tests are performed on each furnace load. One test is run on a sample pin that accompanies each load. The other test is a destructive test, where an actual part is cross-sectioned. This test involves running a carbon analysis on the surface of both the gear pitch (top of the gear tooth) and the gear root (between the gear teeth). The data below are the results of the pitch carbon analysis test for 32 parts.

The regressors are furnace temperature (TEMP), carbon concentration and duration of the carburizing cycle (SOAKPCT, SOAKTIME), and carbon concentration and duration of the diffuse cycle (DIFFPCT, DIFFTIME).

(a) Fit a linear regression model relating the results of the pitch carbon analysis test (PITCH) to the five regressor variables.

(b) Use the model in part (a) to predict PITCH when TEMP = 1650, SOAKTIME = 1.00, SOAKPCT = 1.10, DIFFTIME = 1.00, and DIFFPCT = 0.80.

11-11. Statistics for 21 National Hockey League teams were obtained from the *Hockey Encyclopedia* and are shown in the table on page 503.

TEMP	SOAKTIME	SOAKPCT	DIFFTIME	DIFFPCT	PITCH
1650	0.58	1.10	0.25	0.90	0.013
1650	0.66	1.10	0.33	0.90	0.016
1650	0.66	1.10	0.33	0.90	0.015
1650	0.66	1.10	0.33	0.95	0.016
1600	0.66	1.15	0.33	1.00	0.015
1600	0.66	1.15	0.33	1.00	0.016
1650	1.00	1.10	0.50	0.80	0.014
1650	1.17	1.10	0.58	0.80	0.021
1650	1.17	1.10	0.58	0.80	0.018
1650	1.17	1.10	0.58	0.80	0.019
1650	1.17	1.10	0.58	0.90	0.021
1650	1.17	1.10	0.58	0.90	0.019
1650	1.17	1.15	0.58	0.90	0.021
1650	1.20	1.15	1.10	0.80	0.025
1650	2.00	1.15	1.00	0.80	0.025
1650	2.00	1.10	1.10	0.80	0.026
1650	2.20	1.10	1.10	0.80	0.024
1650	2.20	1.10	1.10	0.80	0.025
1650	2.20	1.15	1.10	0.80	0.024
1650	2.20	1.10	1.10	0.90	0.025
1650	2.20	1.10	1.10	0.90	0.027
1650	2.20	1.10	1.50	0.90	0.026
1650	3.00	1.15	1.50	0.80	0.029
1650	3.00	1.10	1.50	0.70	0.030
1650	3.00	1.10	1.50	0.75	0.028
1650	3.00	1.15	1.66	0.85	0.032
1650	3.33	1.10	1.50	0.80	0.033
1700	4.00	1.10	1.50	0.70	0.039
1650	4.00	1.10	1.50	0.70	0.040
1650	4.00	1.15	1.50	0.85	0.035
1700	12.50	1.00	1.50	0.70	0.056
1700	18.50	1.00	1.50	0.70	0.068

The variables and definitions are as follows:

Wins Number of games won in a season.

Pts Points awarded in a season. Two points for winning a game, one point for losing in overtime, zero points for losing in regular time.

GF Goals for. Total goals scored during the season.

GA Goals against. Goals scored against the team during the season.

PPG Power play goals. Points scored while on power play.

PPcT Power play percentage. The number of power play goals divided by the number of power play opportunities.

SHG Short-handed goals scored during the season.

PPGA Power play goals against.

PKPcT Penalty killing percentage. Measures a team's ability to prevent goals while its opponent is on a power play. Opponent power play goals divided by opponent's opportunities.

SHGA Short-handed goals against.

Fit a multiple linear regression model relating wins to the other variables.

11-12. Consider the linear regression model

$$Y_i = \beta_0' + \beta_1(x_{i1} - \bar{x}_1) + \beta_2(x_{i2} - \bar{x}_2) + \epsilon_i$$

where $\bar{x}_1 = \Sigma x_{i1}/n$ and $\bar{x}_2 = \Sigma x_{i2}/n$.

(a) Write out the least squares normal equations for this model.

(b) Verify that the least squares estimate of the intercept in this model is $\hat{\beta}_0' = \Sigma y_i/n = \bar{y}$.

(c) Suppose that we use $y_i - \bar{y}$ as the response variable in the model above. What effect will this have on the least squares estimate of the intercept?

Team	Wins	Pts	GF	GA	PPG	PPcT	SHG	PPGA	PKPcT	SHGA
Chicago	47	104	338	268	86	27.2	4	71	76.6	6
Minnesota	40	96	321	290	91	26.4	17	67	80.7	20
Toronto	28	68	23	330	79	22.3	13	83	75	9
St. Louis	25	65	285	316	67	21.2	9	63	81.3	12
Detroit	21	57	263	344	37	19.3	7	80	72.6	9
Edmonton	47	106	424	315	86	29.3	22	89	77.5	6
Calgary	32	78	321	317	90	27	7	59	77.1	6
Vancouver	30	75	303	309	65	23.8	5	56	80.8	13
Winnipeg	33	74	311	333	78	23.6	10	67	72.8	7
Los Angeles	27	66	308	365	81	23.8	10	94	68.2	14
Philadelphia	49	106	326	240	60	21.6	15	61	82	7
NY Islanders	42	96	302	226	69	25.8	10	55	83.4	3
Washington	39	94	306	283	75	20.9	3	53	81.6	11
NY Rangers	35	80	306	287	71	22.4	12	75	76	8
New Jersey	17	48	230	338	66	21.9	74	78	73.5	10
Pittsburgh	18	45	257	394	81	22.6	3	110	72.2	15
Boston	50	110	327	228	67	22.2	8	53	80.7	6
Montreal	42	98	350	286	64	22.2	8	68	73.8	8
Buffalo	38	89	318	285	67	21.5	12	48	82.5	9
Quebec	34	80	343	336	61	20.7	6	92	73.6	6
Hartford	19	45	261	403	51	19.3	6	70	76.1	9

11-4 PROPERTIES OF THE LEAST SQUARES ESTIMATORS AND ESTIMATION OF σ^2

The statistical properties of the least squares estimators $\hat{\beta}_0, \hat{\beta}_1, \ldots, \hat{\beta}_k$ may be easily found, under certain assumptions on the error terms $\epsilon_1, \epsilon_2, \ldots, \epsilon_n$, in the regression model. Paralleling the assumptions made in Chapter 10, we assume that the errors ϵ_i are statistically independent with mean zero and variance σ^2. Under these assumptions, the least squares estimators $\hat{\beta}_0, \hat{\beta}_1, \ldots, \hat{\beta}_k$ are *unbiased* estimators of the regression coefficients $\beta_0, \beta_1, \ldots, \beta_k$. This property may be shown as follows:

$$\begin{aligned} E(\hat{\boldsymbol{\beta}}) &= E[(\mathbf{X}'\mathbf{X})^{-1}\mathbf{X}'\mathbf{Y}] \\ &= E[(\mathbf{X}'\mathbf{X})^{-1}\mathbf{X}'(\mathbf{X}\boldsymbol{\beta} + \boldsymbol{\epsilon})] \\ &= E[(\mathbf{X}'\mathbf{X})^{-1}\mathbf{X}'\mathbf{X}\boldsymbol{\beta} + (\mathbf{X}'\mathbf{X})^{-1}\mathbf{X}'\boldsymbol{\epsilon}] \\ &= \boldsymbol{\beta} \end{aligned}$$

since $E(\boldsymbol{\epsilon}) = \mathbf{0}$ and $(\mathbf{X}'\mathbf{X})^{-1}\mathbf{X}'\mathbf{X} = \mathbf{I}$, the identity matrix. Thus, $\hat{\boldsymbol{\beta}}$ is an unbiased estimator of $\boldsymbol{\beta}$.

The variances of the $\hat{\beta}$'s are expressed in terms of the elements of the inverse of the $\mathbf{X}'\mathbf{X}$ matrix. The inverse of $\mathbf{X}'\mathbf{X}$ times the constant σ^2 represents the **covariance matrix** of the regression coefficients $\hat{\boldsymbol{\beta}}$. The diagonal elements of $\sigma^2\,(\mathbf{X}'\mathbf{X})^{-1}$ are the variances of $\hat{\beta}_0, \hat{\beta}_1, \ldots, \hat{\beta}_k$, and the off-diagonal elements of this matrix are the covariances. For example, if we have $k = 2$ regressors, such as in the pull-strength problem, then

$$\begin{aligned} \mathbf{C} &= (\mathbf{X}'\mathbf{X})^{-1} \\ &= \begin{bmatrix} C_{00} & C_{01} & C_{02} \\ C_{10} & C_{11} & C_{12} \\ C_{20} & C_{21} & C_{22} \end{bmatrix} \end{aligned}$$

which is symmetric ($C_{10} = C_{01}$, $C_{20} = C_{02}$, and $C_{21} = C_{12}$) because $(\mathbf{X}'\mathbf{X})^{-1}$ is symmetric, and we have

$$\begin{aligned} V(\hat{\beta}_j) &= \sigma^2 C_{jj}, & j &= 0, 1, 2 \\ \text{cov}(\hat{\beta}_i, \hat{\beta}_j) &= \sigma^2 C_{ij}, & i &\neq j \end{aligned}$$

In general, the covariance matrix of $\hat{\boldsymbol{\beta}}$ is a $(p \times p)$ symmetric matrix whose jjth element is the variance of $\hat{\beta}_j$ and whose i, jth element is the covariance between $\hat{\beta}_i$ and $\hat{\beta}_j$, that is,

$$\text{cov}(\hat{\boldsymbol{\beta}}) = \sigma^2(\mathbf{X}'\mathbf{X})^{-1} = \sigma^2\mathbf{C}$$

The estimates of the variances of these regression coefficients are obtained by replacing σ^2 with an appropriate estimate. When σ^2 is replaced by an estimate $\hat{\sigma}^2$, the square root of the estimated variance of the jth regression coefficient is called the **estimated standard**

error of $\hat{\beta}_j$, or $se(\hat{\beta}_j) = \sqrt{\hat{\sigma}^2 C_{jj}}$. Just as in the simple linear regression case, the estimate of σ^2 is defined in terms of the sum of the squares of the residuals

$$SS_E = \sum_{i=1}^{n} (y_i - \hat{y}_i)^2 = \sum_{i=1}^{n} e_i^2$$

Since it can be shown that the expected value of the quantity SS_E is $\sigma^2(n - p)$, an unbiased estimator of σ^2 is given by the **error** or **residual mean square**

$$\hat{\sigma}^2 = MS_E = \frac{SS_E}{n - p} \tag{11-16}$$

The MS_E will appear in the printout of most multiple regression computer programs. However, it is easy to obtain a computing formula. We can write

$$SS_E = \sum_{i=1}^{n} (y_i - \hat{y}_i)^2 = \sum_{i=1}^{n} e_i^2 = \mathbf{e}'\mathbf{e}$$

Substituting $\mathbf{e} = \mathbf{y} - \hat{\mathbf{y}} = \mathbf{y} - \mathbf{X}\hat{\boldsymbol{\beta}}$, we can show that

$$SS_E = \mathbf{y}'\mathbf{y} - \hat{\boldsymbol{\beta}}'\mathbf{X}'\mathbf{y} \tag{11-17}$$

Therefore, another way to write equation 11-16 is

$$\hat{\sigma}^2 = \frac{SS_E}{n - p} = \frac{\mathbf{y}'\mathbf{y} - \hat{\boldsymbol{\beta}}'\mathbf{X}'\mathbf{y}}{n - p}$$

EXAMPLE 11-3

We will estimate the error variance σ^2 for the multiple regression problem in Examples 11-1 and 11-2. Using the pull strength response data in Table 11-2, we find

$$\mathbf{y}'\mathbf{y} = \sum_{i=1}^{25} y_i^2 = 27{,}177.9510$$

and

$$\hat{\boldsymbol{\beta}}'\mathbf{X}'\mathbf{y} = [2.26379143 \quad 2.74426964 \quad 0.01252781] \begin{bmatrix} 725.82 \\ 8{,}008.37 \\ 274{,}811.31 \end{bmatrix}$$

$$= 27{,}062.7775$$

Therefore, the error sum of squares is found using equation 11-17

$$
\begin{aligned}
SS_E &= \mathbf{y}'\mathbf{y} - \hat{\boldsymbol{\beta}}'\mathbf{X}'\mathbf{y} \\
&= 27{,}177.9510 - 27{,}062.7775 = 115.1735
\end{aligned}
$$

The estimate of σ^2 is computed from equation 11-16 as

$$\hat{\sigma}^2 = \frac{SS_E}{n - p} = \frac{115.1735}{25 - 3} = 5.2352$$

11-5 HYPOTHESIS TESTS IN MULTIPLE LINEAR REGRESSION

In multiple linear regression problems, certain tests of hypotheses about the model parameters are useful in measuring model adequacy. In this section, we describe several important hypothesis-testing procedures. As in the simple linear regression case, hypothesis testing requires that the error terms ϵ_i in the regression model are normally and independently distributed with mean zero and variance σ^2.

11-5.1 Test for Significance of Regression

The test for significance of regression is a test to determine whether a linear relationship exists between the response variable y and a subset of the regressor variables $x_1, x_2, \ldots, x_k$. The appropriate hypotheses are

$$
\begin{aligned}
&H_0\text{: } \beta_1 = \beta_2 = \cdots = \beta_k = 0 \\
&H_1\text{: } \beta_j \neq 0 \quad \text{for at least one } j
\end{aligned}
\tag{11-18}
$$

Rejection of H_0: $\beta_1 = \beta_2 = \cdots = \beta_k = 0$ implies that at least one of the regressor variables $x_1, x_2, \ldots, x_k$ contributes significantly to the model.

The test for significance of regression is a generalization of the procedure used in simple linear regression. The total sum of squares SS_T is partitioned into a sum of squares due to regression and a sum of squares due to error, say,

$$SS_T = SS_R + SS_E$$

and if H_0: $\beta_1 = \beta_2 = \cdots = \beta_k = 0$ is true, then SS_R/σ^2 is a chi-squared random variable with k degrees of freedom. Note that the number of degrees of freedom for this chi-squared random variable is equal to the number of regressor variables in the model. We can also show the SS_E/σ^2 is a chi-squared random variable with $n - p$ degrees of freedom, and that SS_E and SS_R are independent. The test statistic for H_0: $\beta_1 = \beta_2 = \cdots = \beta_k = 0$ is

$$F_0 = \frac{SS_R/k}{SS_E/(n-p)} = \frac{MS_R}{MS_E} \tag{11-19}$$

We should reject H_0 if the computed value of the test statistic in equation 11-19, f_0, is greater than $f_{\alpha,k,n-p}$. The procedure is usually summarized in an analysis of variance table such as Table 11-5.

We have derived a computational formula for SS_E in equation 11-17; that is,

$$SS_E = \mathbf{y}'\mathbf{y} - \hat{\boldsymbol{\beta}}'\mathbf{X}'\mathbf{y}$$

A computational formula for SS_R may be found easily. Now since $SS_T = \sum_{i=1}^n y_i^2 - (\sum_{i=1}^n y_i)^2/n = \mathbf{y}'\mathbf{y} - (\sum_{i=1}^n y_i)^2/n$, we may rewrite the foregoing equation as

$$SS_E = \mathbf{y}'\mathbf{y} - \frac{\left(\sum_{i=1}^n y_i\right)^2}{n} - \left[\hat{\boldsymbol{\beta}}'\mathbf{X}'\mathbf{y} - \frac{\left(\sum_{i=1}^n y_i\right)^2}{n}\right]$$

or

$$SS_E = SS_T - SS_R$$

Therefore, the regression sum of squares is

$$SS_R = \hat{\boldsymbol{\beta}}'\mathbf{X}'\mathbf{y} - \frac{\left(\sum_{i=1}^n y_i\right)^2}{n} \tag{11-20}$$

Table 11-5 Analysis of Variance for Testing Significance of Regression in Multiple Regression

Source of Variation	Sum of Squares	Degrees of Freedom	Mean Square	F_0
Regression	SS_R	k	MS_R	MS_R/MS_E
Error or residual	SS_E	$n - p$	MS_E	
Total	SS_T	$n - 1$		

EXAMPLE 11-4

We will test for significance of regression (with $\alpha = 0.05$) using the wire bond pull strength data from Example 11-1. Some of the numerical quantities required are calculated in Example 11-3. The total sum of squares is

$$SS_T = \mathbf{y}'\mathbf{y} - \frac{\left(\sum_{i=1}^{n} y_i\right)^2}{n}$$
$$= 27{,}177.9510 - \frac{(725.82)^2}{25} = 6105.9447$$

The regression sum of squares is computed from equation 11-20 as follows:

$$SS_R = \hat{\boldsymbol{\beta}}'\mathbf{X}'\mathbf{y} - \frac{\left(\sum_{i=1}^{n} y_i\right)^2}{n}$$
$$= 27{,}062.7775 - \frac{(725.82)^2}{25} = 5990.7712$$

and by subtraction

$$SS_E = SS_T - SS_R$$
$$= \mathbf{y}'\mathbf{y} - \hat{\boldsymbol{\beta}}'\mathbf{X}'\mathbf{y} = 115.1735$$

The analysis of variance is shown in Table 11-6. To test H_0: $\beta_1 = \beta_2 = 0$, we calculate the statistic

$$f_0 = \frac{MS_R}{MS_E} = \frac{2995.3856}{5.2352} = 572.17$$

Since $f_0 > f_{0.05,2,22} = 3.44$ (or since the P-value is considerably smaller than $\alpha = 0.05$), we reject the null hypothesis and conclude that pull strength is linearly related to either wire length or die height, or both. However, we note that this does not necessarily imply that the relationship found is an appropriate model for predicting pull strength as a function of wire length and die height. Further tests of model adequacy are required before we can be comfortable using this model in practice.

Table 11-6 Test for Significance of Regression for Example 11-4

Source of Variation	Sum of Squares	Degrees of Freedom	Mean Square	f_0	P-value
Regression	5990.7712	2	2995.3856	572.17	1.08E-19
Error or residual	115.1735	22	5.2352		
Total	6105.9447	24			

Most multiple regression computer programs provide the test for significance of regression in their output display. The bottom portion of Table 11-4 is the Minitab output for this example. Compare Tables 11-4 and 11-6 and note their equivalence. The P-value is rounded to zero in the computer output.

11-5.2 Tests on Individual Regression Coefficients and Subsets of Coefficients

We are frequently interested in testing hypotheses on the individual regression coefficients. Such tests would be useful in determining the potential value of each of the regressor variables in the regression model. For example, the model might be more effective with the inclusion of additional variables, or perhaps with the deletion of one or more of the regressors presently in the model.

Adding a variable to a regression model always causes the sum of squares for regression to increase and the error sum of squares to decrease. We must decide whether the increase in the regression sum of squares is large enough to justify using the additional variable in the model. Furthermore, adding an unimportant variable to the model can actually increase the error mean square, indicating that adding such a variable has actually made the model a poorer fit to the data.

The hypotheses for testing the significance of any individual regression coefficient, say β_j, are

$$H_0: \beta_j = 0$$
$$H_1: \beta_j \neq 0 \qquad (11\text{-}21)$$

If H_0: $\beta_j = 0$ is not rejected, then this indicates that the regressor x_j can be deleted from the model. The test statistic for this hypothesis is

$$T_0 = \frac{\hat{\beta}_j}{\sqrt{\hat{\sigma}^2 C_{jj}}} \qquad (11\text{-}22)$$

where C_{jj} is the diagonal element of $(\mathbf{X}'\mathbf{X})^{-1}$ corresponding to $\hat{\beta}_j$. Notice that the denominator of equation 11-22 is the standard error of the regression coefficient $\hat{\beta}_j$. The null hypothesis H_0: $\beta_j = 0$ is rejected if $|t_0| > t_{\alpha/2,n-p}$. This is called a **partial** or **marginal test,** because the regression coefficient $\hat{\beta}_j$ depends on all the other regressor variables x_i ($i \neq j$) that are in the model. More will be said about this in the following example.

EXAMPLE 11-5

Consider the wire bond pull strength data, and suppose that we want to test the hypothesis that the regression coefficient for x_2 (die height) is zero. The hypotheses are

$$H_0: \beta_2 = 0$$
$$H_1: \beta_2 \neq 0$$

The main diagonal element of the $(\mathbf{X}'\mathbf{X})^{-1}$ matrix corresponding to $\hat{\beta}_2$ is $C_{22} = 0.0000015$, so the t-statistic in equation 11-24 is

$$t_0 = \frac{\hat{\beta}_2}{\sqrt{\hat{\sigma}^2 C_{22}}} = \frac{0.01253}{\sqrt{(5.2352)(0.0000015)}} = 4.4767$$

Since $t_{0.025,22} = 2.074$, we reject H_0: $\beta_2 = 0$ and conclude that the variable x_2 (die height) contributes significantly to the model. We could also have used a P-value to draw conclusions. The P-value for $t_0 = 4.4767$ is $P = 0.0002$, so with $\alpha = 0.05$ we would reject the null hypothesis. Note that this test measures the marginal or partial contribution of x_2 given that x_1 is in the model. That is, the t-test measures the contribution of adding the variable x_2 = die height to a model that already contains x_1 = wire length. Table 11-4 shows the value of the t-test computed by Minitab. Note that the computer produces a t-test for each regression coefficient in the model. These t-tests indicate that both regressors contribute to the model.

There is another way to test the contribution of an individual regressor variable to the model. This approach determines the increase in the regression sum of squares obtained by adding a variable x_j (say) to the model, given that other variables x_i $(i \neq j)$ are already included in the regression equation.

The procedure used to do this is called the **general regression significance test,** or the **extra sum of squares** method. This procedure can also be used to investigate the contribution of a *subset* of the regressor variables to the model. Consider the regression model with k regressor variables

$$\mathbf{y} = \mathbf{X}\boldsymbol{\beta} + \boldsymbol{\epsilon} \tag{11-23}$$

where $\mathbf{y}$ is $(n \times 1)$, $\mathbf{X}$ is $(n \times p)$, $\boldsymbol{\beta}$ is $(p \times 1)$, $\boldsymbol{\epsilon}$ is $(n \times 1)$, and $p = k + 1$. We would like to determine if the subset of regressor variables $x_1, x_2, \ldots, x_r$ $(r < k)$ as a whole contributes significantly to the regression model. Let the vector of regression coefficients be partitioned as follows:

$$\boldsymbol{\beta} = \begin{bmatrix} \boldsymbol{\beta}_1 \\ \boldsymbol{\beta}_2 \end{bmatrix} \tag{11-24}$$

where $\boldsymbol{\beta}_1$ is $(r \times 1)$ and $\boldsymbol{\beta}_2$ is $[(p - r) \times 1]$. We wish to test the hypotheses

$$H_0: \boldsymbol{\beta}_1 = \mathbf{0}$$
$$H_1: \boldsymbol{\beta}_1 \neq \mathbf{0} \tag{11-25}$$

where $\mathbf{0}$ denotes a vector of zeroes. The model may be written as

$$\mathbf{y} = \mathbf{X}\boldsymbol{\beta} + \boldsymbol{\epsilon} = \mathbf{X}_1\boldsymbol{\beta}_1 + \mathbf{X}_2\boldsymbol{\beta}_2 + \boldsymbol{\epsilon} \tag{11-26}$$

where $\mathbf{X}_1$ represents the columns of $\mathbf{X}$ associated with $\boldsymbol{\beta}_1$ and $\mathbf{X}_2$ represents the columns of $\mathbf{X}$ associated with $\boldsymbol{\beta}_2$.

For the **full model** (including both $\boldsymbol{\beta}_1$ and $\boldsymbol{\beta}_2$), we know that $\hat{\boldsymbol{\beta}} = (\mathbf{X}'\mathbf{X})^{-1}\mathbf{X}'\mathbf{y}$. In addition, the regression sum of squares for all variables including the intercept is

$$SS_R(\boldsymbol{\beta}) = \hat{\boldsymbol{\beta}}'\mathbf{X}'\mathbf{y} \qquad (p = k + 1 \text{ degrees of freedom})$$

and

$$MS_E = \frac{\mathbf{y}'\mathbf{y} - \hat{\boldsymbol{\beta}}\mathbf{X}'\mathbf{y}}{n - p}$$

$SS_R(\boldsymbol{\beta})$ is called the regression sum of squares due to $\boldsymbol{\beta}$. To find the contribution of the terms in $\boldsymbol{\beta}_1$ to the regression, fit the model assuming the null hypothesis H_0: $\boldsymbol{\beta}_1 = \mathbf{0}$ to be true. The **reduced model** is found from equation 11-26 as

$$\mathbf{y} = \mathbf{X}_2\boldsymbol{\beta}_2 + \boldsymbol{\epsilon} \tag{11-27}$$

The least squares estimate of $\boldsymbol{\beta}_2$ is $\hat{\boldsymbol{\beta}}_2 = (\mathbf{X}_2'\mathbf{X}_2)^{-1}\mathbf{X}_2'\mathbf{y}$, and

$$SS_R(\boldsymbol{\beta}_2) = \hat{\boldsymbol{\beta}}_2'\mathbf{X}_2'\mathbf{y} \qquad (p - r \text{ degrees of freedom}) \tag{11-28}$$

The regression sum of squares due to $\boldsymbol{\beta}_1$ given that $\boldsymbol{\beta}_2$ is already in the model is

$$SS_R(\boldsymbol{\beta}_1 \mid \boldsymbol{\beta}_2) = SS_R(\boldsymbol{\beta}) - SS_R(\boldsymbol{\beta}_2) \tag{11-29}$$

This sum of squares has r degrees of freedom. It is sometimes called the extra sum of squares due to $\boldsymbol{\beta}_1$. Note that $SS_R(\boldsymbol{\beta}_1 \mid \boldsymbol{\beta}_2)$ is the increase in the regression sum of squares due to including the variables $x_1, x_2, \ldots, x_r$ in the model. Now $SS_R(\boldsymbol{\beta}_1 \mid \boldsymbol{\beta}_2)$ is independent of MS_E, and the null hypothesis $\boldsymbol{\beta}_1 = \mathbf{0}$ may be tested by the statistic

$$F_0 = \frac{SS_R(\boldsymbol{\beta}_1 \mid \boldsymbol{\beta}_2)/r}{MS_E} \tag{11-30}$$

If the computed value of the test statistic $f_0 > f_{\alpha,r,n-p}$, we reject H_0, concluding that at least one of the parameters in $\boldsymbol{\beta}_1$ is not zero and, consequently, at least one of the variables $x_1, x_2, \ldots, x_r$ in $\mathbf{X}_1$ contributes significantly to the regression model. Some authors call the test in equation 11-30 a **partial *F*-test.**

The partial *F*-test is very useful. We can use it to measure the contribution of each individual regressor x_j as if it were the last variable added to the model by computing

$$SS_R(\beta_j \mid \beta_0, \beta_1, \ldots, \beta_{j-1}, \beta_{j+1}, \ldots, \beta_k), \qquad j = 1, 2, \ldots, k$$

This is the increase in the regression sum of squares due to adding x_j to a model that already includes $x_1, \ldots, x_{j-1}, x_{j+1}, \ldots, x_k$. The partial *F*-test is a more general procedure in that we can measure the effect of sets of variables. In Section 11-10 we show how the partial *F*-test plays a major role in *model building*—that is, in searching for the best set of regressor variables to use in the model.

EXAMPLE 11-6

Consider the wire bond pull strength data in Example 11-1. We will investigate the contribution of the variable x_2 (die height) to the model using the partial *F*-test approach. That is, we wish to test

$$H_0: \beta_2 = 0$$
$$H_1: \beta_2 \neq 0$$

To test this hypothesis, we need the extra sum of squares due to β_2, or

$$\begin{aligned} SS_R(\beta_2 \mid \beta_1, \beta_0) &= SS_R(\beta_1, \beta_2, \beta_0) - SS_R(\beta_1, \beta_0) \\ &= SS_R(\beta_1, \beta_2 \mid \beta_0) - SS_R(\beta_1 \mid \beta_0) \end{aligned}$$

In Example 11-4 we have calculated

$$SS_R(\beta_1, \beta_2 \mid \beta_0) = \hat{\boldsymbol{\beta}}' \mathbf{X}'\mathbf{y} - \frac{\left(\sum_{i=1}^{n} y_i\right)}{n} = 5990.7712 \quad \text{(2 degrees of freedom)}$$

and from Example 10-10, where we fit the model $Y = \beta_0 + \beta_1 x_1 + \epsilon$, we can calculate

$$\begin{aligned} SS_R(\beta_1 \mid \beta_0) &= \hat{\beta}_1 S_{xy} = (2.9027)(2027.7132) \\ &= 5885.8521 \quad \text{(1 degree of freedom)} \end{aligned}$$

Therefore,

$$\begin{aligned} SS_R(\beta_2 \mid \beta_1, \beta_0) &= 5990.7712 - 5885.8521 \\ &= 104.9191 \quad \text{(1 degree of freedom)} \end{aligned}$$

This is the increase in the regression sum of squares due to adding x_2 to a model already containing x_1. To test H_0: $\beta_2 = 0$, calculate the test statistic

$$f_0 = \frac{SS_R(\beta_2 \mid \beta_1, \beta_0)/1}{MS_E} = \frac{104.9191/1}{5.2352} = 20.04$$

Note that the MS_E from the full model, using both x_1 and x_2, is used in the denominator of the test statistic. Since $f_{0.05,1,22} = 4.30$, we reject H_0: $\beta_2 = 0$ and conclude that the regressor die height (x_2) contributes significantly to the model.

Table 11-4 shows the Minitab regression output for the wire bond pull strength data. At the bottom of this table the quantity labeled "Seq SS" shows the sum of squares obtained by fitting x_1 alone (5885.9) and the sum of squares obtained by fitting x_2 after x_1. Notationally, these are referred to above as $SS_R(\beta_1 \mid \beta_0)$ and $SS_R(\beta_2 \mid \beta_1, \beta_0)$.

Since the partial F-test in the above example involves a single variable, it is equivalent to the t-test. To see this, recall from Example 11-6 that the t-test on H_0: $\beta_2 = 0$ resulted in the test statistic $t_0 = 4.4767$. Furthermore, the square of a t-random variable with υ degrees of freedom is an F-random variable with one and υ degrees of freedom, and we note that $t_0^2 = (4.4767)^2 = 20.04 = f_0$.

EXERCISES FOR SECTIONS 11-4 AND 11-5

11-13. Consider the regression model fit to the soil shear strength data in Exercise 11-1.
(a) Test for significance of regression using $\alpha = 0.05$. What is the P-value for this test?
(b) Estimate σ^2.
(c) Estimate the standard errors of the regression coefficients $\hat{\beta}_1$ and $\hat{\beta}_2$.
(d) Construct the t-test on each regression coefficient. What are your conclusions, using $\alpha = 0.05$?

11-14. Consider the absorption index data in Exercise 11-2. The total sum of squares for y is $S_{yy} = 742.00$.
(a) Test for significance of regression using $\alpha = 0.01$. What is the P-value for this test?
(b) Estimate σ^2.
(c) Estimate the standard error of $\hat{\beta}_1$.
(d) Test the hypothesis H_0: $\beta_1 = 0$ versus H_1: $\beta_1 \neq 0$ using $\alpha = 0.01$. What is the P-value for this test? What conclusion can you draw about the usefulness of x_1 as a regressor in this model?

11-15. Consider the NFL data in Exercise 11-4.
(a) Test for significance of regression using $\alpha = 0.05$. What is the P-value for this test?
(b) Estimate σ^2.
(c) Estimate the standard error of $\hat{\beta}_2$.
(d) Conduct the t-test for each regression coefficient β_2, β_7, and β_8. Using $\alpha = 0.05$, what conclusions can you draw about the variables in this model?

11-16. Reconsider the NFL data in Exercise 11-4.
(a) Find the amount by which the regressor x_8 (opponents' yards rushing) increases the regression sum of squares.
(b) Use the results from part (a) above and Exercise 11-15 to conduct an

F-test for H_0: $\beta_8 = 0$ versus H_1: $\beta_8 \neq 0$ using $\alpha = 0.05$. What is the P-value for this test? What conclusions can you draw?

11-17. Consider the gasoline mileage data in Exercise 11-5.

(a) Test for significance of regression using $\alpha = 0.05$. What conclusions can you draw?

(b) Find the t-test statistic for both regressors. Using $\alpha = 0.05$, what conclusions can you draw? Do both regressors contribute to the model?

11-18. A regression model $Y = \beta_0 + \beta_1 x_1 + \beta_2 x_2 + \beta_3 x_3 + \epsilon$ has been fit to a sample of $n = 25$ observations. The calculated t-ratios $\hat{\beta}_j / se(\hat{\beta}_j)$, $j = 1, 2, 3$ are as follows: for β_1, $t_0 = 4.82$, for β_2, $t_0 = 8.21$, and for β_3, $t_0 = 0.98$.

(a) Find P-values for each of the t-statistics.

(b) Using $\alpha = 0.05$, what conclusions can you draw about the regressor x_3? Does it seem likely that this regressor contributes significantly to the model?

11-19. Consider the electric power consumption data in Exercise 11-6.

(a) Test for significance of regression using $\alpha = 0.01$. What is the P-value for this test?

(b) Estimate σ^2.

(c) Use the t-test to assess the contribution of each regressor to the model. Using $\alpha = 0.01$, what conclusions can you draw?

11-20. Consider the bearing wear data in Exercise 11-7 with no interaction.

(a) Test for significance of regression using $\alpha = 0.05$. What is the P-value for this test? What are your conclusions?

(b) Compute the t-statistics for each regression coefficient. Using $\alpha = 0.05$, what conclusions can you draw?

(c) Use the extra sum of squares method to investigate the usefulness of adding x_2 = load to a model that already contains x_1 = oil viscosity. Use $\alpha = 0.05$.

11-21. Reconsider the bearing wear data from Exercises 11-7 and 11-20.

(a) Refit the model with an interaction term. Test for significance of regression using $\alpha = 0.05$.

(b) Use the extra sum of squares method to determine whether the interaction term contributes significantly to the model. Use $\alpha = 0.05$.

(c) Estimate σ^2 for the interaction model. Compare this to the estimate of σ^2 from the model in Exercise 11-20.

11-22. Consider the wire bond pull strength data in Exercise 11-8.

(a) Test for significance of regression using $\alpha = 0.05$. Find the P-value for this test. What conclusions can you draw?

(b) Estimate σ^2 for this model.

(c) Calculate the t-test statistic for each regression coefficient. Using $\alpha = 0.05$, what conclusions can you draw? Do all variables contribute to the model?

11-23. Reconsider the semiconductor data in Exercise 11-9.

(a) Test for significance of regression using $\alpha = 0.05$. What conclusions can you draw?

(b) Estimate σ^2 for this model.

(c) Calculate the t-test statistic for each regression coefficient. Using $\alpha = 0.05$, what conclusions can you draw?

11-24. Exercise 11-10 presents data on heat treating gears.

(a) Test the regression model for significance of regression. Using $\alpha = 0.05$, find the P-value for the test and draw conclusions.

(b) Estimate σ^2 for this model.

(c) Evaluate the contribution of each regressor to the model using the t-test with $\alpha = 0.05$.

(d) Fit a new model to the response PITCH using new regressors x_1 = SOAKTIME $\times$ SOAKPCT and x_2 = DIFFTIME $\times$ DIFFPCT.

(e) Test the model in part (d) for significance of regression using $\alpha = 0.05$. Also calculate the t-test for each regressor and draw conclusions.

(f) Estimate σ^2 for the model from part (d) and compare this to the estimate of σ^2 obtained in part (b) above. Which estimate is smaller? Does this offer any insight regarding which model might be preferable?

11-25. Data on National Hockey League team performance was presented in Exercise 11-11.

(a) Test the model from this exercise for significance of regression using $\alpha = 0.05$. What conclusions can you draw?

(b) Use the t-test to evaluate the contribution of each regressor to the model. Does it seem that all regressors are necessary? Use $\alpha = 0.05$.

(c) Fit a regression model relating the number of games won to the number of points scored and the number of power play goals. Does this seem to be a logical choice of regressors, considering your answer to part (b) above? Test this new model for significance of regression and evaluate the contribution of each regressor to the model using the t-test. Use $\alpha = 0.05$.

11-6 CONFIDENCE INTERVALS IN MULTIPLE LINEAR REGRESSION

11-6.1 Confidence Intervals on Individual Regression Coefficients

In multiple regression models, it is often useful to construct confidence interval estimates for the regression coefficients $\{\beta_j\}$. The development of a procedure for obtaining these confidence intervals requires that the errors $\{\epsilon_i\}$ are normally and independently distributed with mean zero and variance σ^2. This is the same assumption required in hypothesis testing. Therefore, the observations $\{Y_i\}$ are normally and independently distributed with mean $\beta_0 + \sum_{j=1}^{k} \beta_j x_{ij}$ and variance σ^2. Since the least squares estimator $\hat{\boldsymbol{\beta}}$ is a linear combination of the observations, it follows that $\hat{\boldsymbol{\beta}}$ is normally distributed with mean vector $\boldsymbol{\beta}$ and covariance matrix $\sigma^2(\mathbf{X}'\mathbf{X})^{-1}$. Then each of the statistics

$$\frac{\hat{\beta}_j - \beta_j}{\sqrt{\hat{\sigma}^2 C_{jj}}} \qquad j = 0, 1, \ldots, k \tag{11-31}$$

has a t distribution with $n - p$ degrees of freedom, where C_{jj} is the jjth element of the $(\mathbf{X}'\mathbf{X})^{-1}$ matrix, and $\hat{\sigma}^2$ is the estimate of the error variance, obtained from equation 11-16. This leads to the following $100(1 - \alpha)$ percent confidence interval for the regression coefficient β_j, $j = 0, 1, \ldots, k$.

Definition

A 100(1 − α) percent **confidence interval on the regression coefficient** β_j, $j = 0, 1, \ldots, k$ in the multiple linear regression model is given by

$$\hat{\beta}_j - t_{\alpha/2, n-p}\sqrt{\hat{\sigma}^2 C_{jj}} \leq \beta_j \leq \hat{\beta}_j + t_{\alpha/2, n-p}\sqrt{\hat{\sigma}^2 C_{jj}} \qquad (11\text{-}32)$$

EXAMPLE 11-7

We will construct a 95% confidence interval on the parameter β_1 in the wire bond pull strength problem. The point estimate of β_1 is $\hat{\beta}_1 = 2.74427$ and the diagonal element of $(\mathbf{X}'\mathbf{X})^{-1}$ corresponding to β_1 is $C_{11} = 0.001671$. The estimate of σ^2 was obtained in Example 11-3 as 5.2352, and $t_{0.025,22} = 2.074$. Therefore, the 95% confidence interval on β_1 is computed from equation 11-32 as

$$2.74427 - (2.074)\sqrt{(5.2352)(.001671)} \leq \beta_1 \leq 2.74427 + (2.074)\sqrt{(5.2352)(.001671)}$$

which reduces to

$$2.55029 \leq \beta_1 \leq 2.93825$$

11-6.2 Confidence Interval on the Mean Response

We may also obtain a confidence interval on the mean response at a particular point, say, $x_{01}, x_{02}, \ldots, x_{0k}$. To estimate the mean response at this point, define the vector

$$\mathbf{x}_0 = \begin{bmatrix} 1 \\ x_{01} \\ x_{02} \\ \vdots \\ x_{0k} \end{bmatrix}$$

The mean response at this point is $E(Y \mid \mathbf{x}_0) = \mu_{Y|\mathbf{x}_0} = \mathbf{x}_0'\boldsymbol{\beta}$, which is estimated by

$$\hat{\mu}_{Y|\mathbf{x}_0} = \mathbf{x}_0'\hat{\boldsymbol{\beta}} \qquad (11\text{-}33)$$

This estimator is unbiased, since $E(\mathbf{x}_0'\hat{\boldsymbol{\beta}}) = \mathbf{x}_0'\boldsymbol{\beta} = E(Y \mid \mathbf{x}_0) = \mu_{Y|\mathbf{x}_0}$ and the variance of $\hat{\mu}_{Y|\mathbf{x}_0}$ is

$$V(\hat{\mu}_{Y|\mathbf{x}_0}) = \sigma^2 \mathbf{x}_0'(\mathbf{X}'\mathbf{X})^{-1}\mathbf{x}_0 \qquad (11\text{-}34)$$

A 100(1 − α) percent confidence interval on $\mu_{Y|\mathbf{x}_0}$ can be constructed from the statistic

$$\frac{\hat{\mu}_{Y|\mathbf{x}_0} - \mu_{Y|\mathbf{x}_0}}{\sqrt{\hat{\sigma}^2 \mathbf{x}_0'(\mathbf{X}'\mathbf{X})^{-1}\mathbf{x}_0}} \tag{11-35}$$

Definition

For the multiple linear regression model, a 100(1 − α) percent **confidence interval on the mean response** at the point $x_{01}, x_{02}, \ldots, x_{0k}$ is

$$\hat{\mu}_{Y|\mathbf{x}_0} - t_{\alpha/2,n-p}\sqrt{\hat{\sigma}^2 \mathbf{x}_0'(\mathbf{X}'\mathbf{X})^{-1}\mathbf{x}_0} \le \mu_{Y|\mathbf{x}_0} \le \hat{\mu}_{Y|\mathbf{x}_0} + t_{\alpha/2,n-p}\sqrt{\hat{\sigma}^2 \mathbf{x}_0'(\mathbf{X}'\mathbf{X})^{-1}\mathbf{x}_0} \tag{11-36}$$

Equation 11-36 is a confidence interval about the regression hyperplane. It is the multiple regression generalization of equation 10-33.

EXAMPLE 11-8

The engineer in Example 11-1 would like to construct a 95% confidence interval on the mean pull strength for a wire bond with wire length $x_1 = 8$ and die height $x_2 = 275$. Therefore,

$$\mathbf{x}_0 = \begin{bmatrix} 1 \\ 8 \\ 275 \end{bmatrix}$$

The estimated mean response at this point is found from equation 11-33 as

$$\hat{\mu}_{Y|\mathbf{x}_0} = \mathbf{x}_0'\hat{\boldsymbol{\beta}} = [1 \quad 8 \quad 275]\begin{bmatrix} 2.26379 \\ 2.74427 \\ 0.01253 \end{bmatrix} = 27.66$$

The variance of $\hat{\mu}_{Y|\mathbf{x}_0}$ is estimated by

$$\hat{\sigma}^2 \mathbf{x}_0'(\mathbf{X}'\mathbf{X})^{-1}\mathbf{x}_0 = 5.2352\,[1\ 8\ 275]$$

$$\times \begin{bmatrix} .214653 & -.007491 & -.000340 \\ -.007491 & .001671 & -.000019 \\ -.000340 & -.000019 & .0000015 \end{bmatrix}\begin{bmatrix} 1 \\ 8 \\ 275 \end{bmatrix}$$

$$= 5.2352\,(0.04444) = 0.23266$$

Therefore, a 95% confidence interval on the mean pull strength at this point is found from equation 11-36 as

$$27.66 - 2.074\sqrt{0.23266} \leq \mu_{Y|\mathbf{x}_0} \leq 27.66 + 2.074\sqrt{0.23266}$$

which reduces to

$$26.66 \leq \mu_{Y|\mathbf{x}_0} \leq 28.66$$

11-7 PREDICTION OF NEW OBSERVATIONS

A regression model can be used to predict future observations on the response variable Y corresponding to particular values of the independent variables, say, $x_{01}, x_{02}, \ldots, x_{0k}$. If $\mathbf{x}_0' = [1, x_{01}, x_{02}, \ldots, x_{0k}]$, then a point estimate of the future observation Y_0 at the point $x_{01}, x_{02}, \ldots, x_{0k}$ is

$$\hat{y}_0 = \mathbf{x}_0'\hat{\boldsymbol{\beta}} \tag{11-37}$$

A $100(1 - \alpha)$ percent **prediction interval for this future observation** is

$$\hat{y}_0 - t_{\alpha/2,n-p}\sqrt{\hat{\sigma}^2(1 + \mathbf{x}_0'(\mathbf{X}'\mathbf{X})^{-1}\mathbf{x}_0)} \leq y_0 \leq \hat{y}_0 + t_{\alpha/2,n-p}\sqrt{\hat{\sigma}^2(1 + \mathbf{x}_0'(\mathbf{X}'\mathbf{X})^{-1}\mathbf{x}_0)} \tag{11-38}$$

This prediction interval is a generalization of the prediction interval given in equation 10-35 for a future observation in simple linear regression. If you compare the prediction interval equation 11-38 above with the expression for the confidence interval on the mean, equation 11-36, you will observe that the prediction interval is always wider than the confidence interval. The confidence interval expresses the error in estimating the mean of a distribution, while the prediction interval expresses the error in predicting a future observation from the distribution at the point $\mathbf{x}_0$. This must include the error in estimating the mean at that point, as well as the inherent variability in the random variable Y at the same value $\mathbf{x} = \mathbf{x}_0$.

In predicting new observations and in estimating the mean response at a given point $x_{01}, x_{02}, \ldots, x_{0k}$, we must be careful about extrapolating beyond the region containing the original observations. It is very possible that a model that fits well in the region of the original data will no longer fit well outside of that region. In multiple regression it is

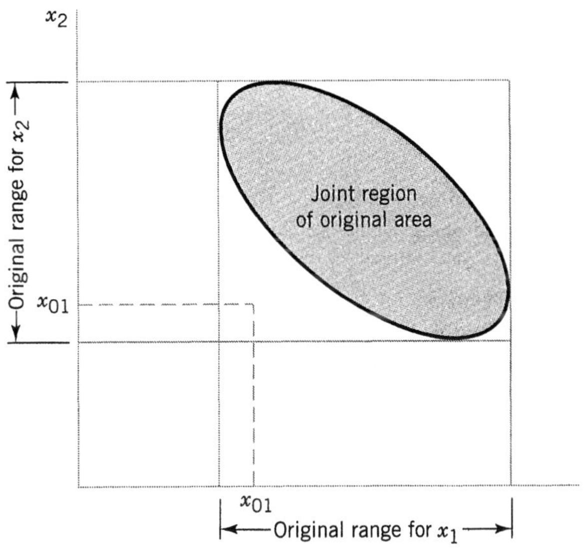

Figure 11-4 An example of extrapolation in multiple regression.

often easy to inadvertently extrapolate, since the levels of the variables $(x_{i1}, x_{i2}, \ldots, x_{ik})$, $i = 1, 2, \ldots, n$, jointly define the region containing the data. As an example, consider Fig. 11-4, which illustrates the region containing the observations for a two-variable regression model. Note that the point (x_{01}, x_{02}) lies within the ranges of both regressor variables x_1 and x_2, but it is outside the region that is actually spanned by the original observations. Thus, either predicting the value of a new observation or estimating the mean response at this point is an extrapolation of the original regression model.

EXAMPLE 11-9

Suppose that the engineer in Example 11-1 wishes to construct a 95% prediction interval on the wire bond pull strength when the wire length is $x_1 = 8$ and the die height is $x_2 = 275$. Note that $\mathbf{x}_0' = [1 \quad 8 \quad 275]$, and the point estimate of the pull strength is $\hat{y}_0 = \mathbf{x}_0'\hat{\boldsymbol{\beta}} = 27.66$. Also, in Example 11-8 we calculated $\mathbf{x}_0'(\mathbf{X}'\mathbf{X})^{-1}\mathbf{x}_0 = 0.04444$. Therefore, from equation 11-38 we have

$$27.66 - 2.074\sqrt{5.2352(1 + 0.04444)} \leq y_0 \leq 27.66 + 2.074\sqrt{5.2352(1 + 0.04444)}$$

and the 95% prediction interval is

$$22.81 \leq y_0 \leq 32.51$$

Notice that the prediction interval is wider than the confidence interval on the mean response at the same point, calculated in Example 11-8.

EXERCISES FOR SECTIONS 11-6 AND 11-7

11-26. Consider the soil absorption data in Exercise 11-2.

(a) Find a 95% confidence interval on the regression coefficient β_1.

(b) Find a 95% confidence interval on mean soil absorption index when $x_1 = 200$ and $x_2 = 50$.

(c) Find a 95% prediction interval on the soil absorption index when $x_1 = 200$ and $x_2 = 50$.

11-27. Consider the NFL data in Exercise 11-4.

(a) Find a 95% confidence interval on β_8.

(b) What is the estimated standard error of $\hat{\mu}_{Y|\mathbf{x}_0}$ when $x_2 = 2000$ yards, $x_7 = 60\%$, and $x_8 = 1800$ yards?

(c) Find a 95% confidence interval on the mean number of games won when $x_2 = 2000$, $x_7 = 60$, and $x_8 = 1800$.

11-28. Consider the gasoline mileage data in Exercise 11-5.

(a) Find 99% confidence intervals on β_1 and β_6.

(b) Find a 99% confidence interval on the mean of Y when $x_1 = 300$ and $x_6 = 4$.

(c) Fit a new regression model to these data using x_1, x_2, x_6, and x_{10} as the regressors. Find 99% confidence intervals on the regression coefficients in this new model.

(d) Compare the lengths of the confidence intervals on β_1 and β_6 from part (c) above with those found in part (a). Which intervals are longer? Does this offer any insight about adding the variables x_2 and x_{10} to the model?

11-29. Consider the electric power consumption data in Exercise 11-6.

(a) Find 95% confidence intervals on β_1, β_2, β_3, and β_4.

(b) Find a 95% confidence interval on the mean of Y when $x_1 = 75$, $x_2 = 24$, $x_3 = 90$, and $x_4 = 98$.

(c) Find a 95% prediction interval on the power consumption when $x_1 = 75$, $x_2 = 24$, $x_3 = 90$, and $x_4 = 98$.

11-30. Consider the bearing wear data in Exercise 11-7.

(a) Find 99% confidence intervals on β_1 and β_2.

(b) Recompute the confidence intervals in part (a) after the interaction term x_1x_2 is added to the model. Compare the lengths of these confidence intervals with those computed in part (a). Do the lengths of these intervals provide any information about the contribution of the interaction term in the model?

11-31. Consider the wire bond pull strength data in Exercise 11-8.

(a) Find 95% confidence intervals on the regression coefficients.

(b) Find a 95% confidence interval on mean pull strength when $x_2 = 20$, $x_3 = 30$, $x_4 = 90$ and $x_5 = 2.0$.

(c) Find a 95% prediction interval on pull strength when $x_2 = 20$, $x_3 = 30$, $x_4 = 90$, and $x_5 = 2.0$.

11-32. Consider the semiconductor data in Exercise 11-9.

(a) Find 99% confidence intervals on the regression coefficients.

(b) Find a 99% prediction interval on HFE when $x_1 = 14.5$, $x_2 = 220$, and $x_3 = 5.0$.

(c) Find a 99% confidence interval on mean HFE when $x_1 = 14.5$, $x_2 = 220$, and $x_3 = 5.0$.

11-33. Consider the heat treating data from Exercise 11-10.

(a) Find 95% confidence intervals on the regression coefficients.

(b) Find a 95% confidence interval on mean PITCH when TEMP = 1650, SOAKTIME = 1.00, SOAKPCT = 1.10, DIFFTIME = 1.00, and DIFFPCT = 0.80.

11-34. Reconsider the heat treating data in Exercises 11-10 and 11-24, where we fit a model to PITCH using regressors x_1 = SOAKTIME × SOAKPCT and x_2 = DIFFTIME × DIFFPCT.

(a) Using the model with regressors x_1 and x_2, find a 95% confidence interval on mean PITCH when SOAKTIME = 1.00, SOAKPCT = 1.10, DIFFTIME = 1.00, and DIFFPCT = 0.80.

(b) Compare the length of this confidence interval with the length of the confidence interval on mean PITCH at the same point from Exercise 11-33 part (b), where an additive model in SOAKTIME, SOAKPCT, DIFFTIME, and DIFFPCT was used. Which confidence interval is shorter? Does this tell you anything about which model is preferable?

11-35. Consider the NHL data in Exercise 11-11.

(a) Find a 95% confidence interval on the regression coefficient for the variable "Pts."

(b) Fit a simple linear regression model relating the response variable "wins" to the regressor "Pts."

(c) Find a 95% confidence interval on the slope for the simple linear regression model from part (b) above.

(d) Compare the lengths of the two confidence intervals computed in parts (a) and (c) above. Which interval is shorter? Does this tell you anything about which model is preferable?

11-8 MEASURES OF MODEL ADEQUACY

A number of techniques can be used to measure the adequacy of a multiple regression model. This section will present several of these techniques.

11-8.1 Coefficient of Multiple Determination (R^2)

The coefficient of multiple determination R^2 is defined as

$$R^2 = \frac{SS_R}{SS_T} = 1 - \frac{SS_E}{SS_T} \tag{11-39}$$

R^2 is a measure of the amount of reduction in the variability of y obtained by using the regressor variables $x_1, x_2, \ldots, x_k$. As in the simple linear regression case, we must have $0 \leq R^2 \leq 1$. However, a large value of R^2 does not necessarily imply that the regression model is a good one. Adding a variable to the model will always increase R^2, regardless of whether or not the additional variable is statistically significant. Thus, models that have large values of R^2 may yield poor predictions of new observations or estimates of the mean response.

The positive square root of R^2 is called the **multiple correlation coefficient** between y and the set of regressor variables $x_1, x_2, \ldots, x_k$. That is, R is a measure of the linear association between y and $x_1, x_2, \ldots, x_k$. When $k = 1$, this becomes the simple correlation between y and x.

EXAMPLE 11-10

The coefficient of multiple determination for the regression model fit to the wire bond pull strength data in Example 11-1 is

$$R^2 = \frac{SS_R}{SS_T} = \frac{5990.7712}{6105.9447} = 0.981137$$

That is, about 98.11% of the variability in pull strength y has been explained when the two regressor variables, wire length (x_1) and die height (x_2), are used. In Example 10-10, a simple linear regression model relating y to x_1 was developed. The value of R^2 for this model is $R^2 = 0.963954$. Therefore, adding the variable x_2 to the model has increased R^2 from 0.963954 to 0.981137.

11-8.2 Residual Analysis

The residuals from the multiple regression model, defined by $e_i = y_i - \hat{y}_i$, play an important role in judging model adequacy just as they do in simple linear regression. As noted in Section 10-8.1, several residual plots are often useful; these are illustrated in Example 11-11. It is also helpful to plot the residuals against variables not presently in the model that are possible candidates for inclusion. Patterns in these plots, similar to those in Fig. 10-7, indicate that the model may be improved by adding the candidate variable.

EXAMPLE 11-11

The residuals for the model from Example 11-1 are shown in Table 11-3. A normal probability plot of these residuals is shown in Fig. 11-5. No severe deviations from normality are obviously apparent, although the two largest residuals ($e_{15} = 5.88$ and $e_{17} = 4.33$) do not fall extremely close to a straight line drawn through the remaining residuals.

The **standardized residuals**

$$d_i = \frac{e_i}{\sqrt{MS_E}} = \frac{e_i}{\sqrt{\hat{\sigma}^2}}$$

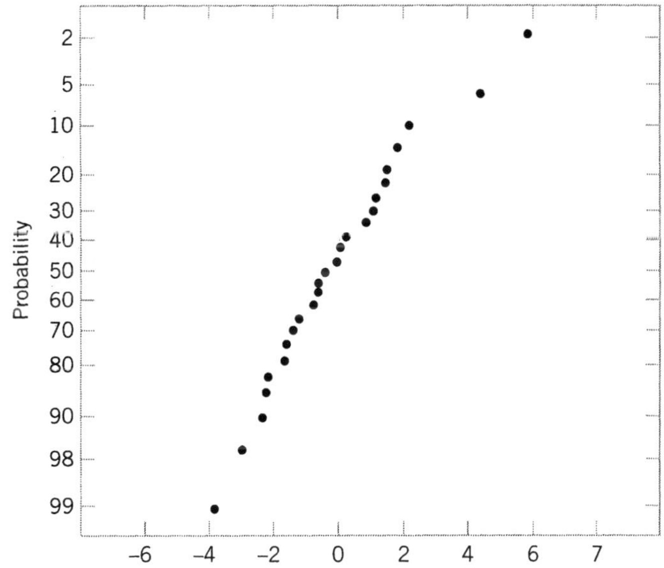

Figure 11-5 Normal probability plot of residuals.

are often more useful than the ordinary residuals when assessing residual magnitude. The standardized residuals corresponding to e_{15} and e_{17} are $d_{15} = 5.88/\sqrt{5.2352} = 2.57$ and $d_{17} = 4.33/\sqrt{4.2352} = 1.89$, and they do not seem unusually large. Inspection of the data does not reveal any error in collecting observations 15 and 17, nor does it produce any other reason to discard or modify these two points.

The residuals are plotted against $\hat{y}$ in Fig. 11-6, and against x_1 and x_2 in Figs. 11-7 and 11-8, respectively.[1] The two largest residuals, e_{15} and e_{17}, are apparent. Figure 11-7 gives some indication that the model underpredicts the pull strength for assemblies with short wire length ($x_1 \leq 6$) and long wire length ($x_1 \geq 15$), and overpredicts the strength for assemblies with intermediate wire length ($7 \leq x_1 \leq 14$). The same impression is obtained from Fig. 11-6. Either the relationship between strength and wire length is linear (requiring that a term involving x_1^2, say, be added to the model), or other regressor variables not presently in the model affected the response. Subsequently, we will see that a third regressor variable is required to adequately model these data.

In Example 11-11 we used the standardized residuals $d_i = e_i/\sqrt{\hat{\sigma}^2}$ as a measure of residual magnitude. Some analysts prefer to plot standardized residuals instead of ordinary residuals, because the standardized residuals are scaled so that their standard deviation is approximately unity. Consequently, large residuals (that may indicate possible outliers or unusual observations) will be more obvious from inspection of the residual plots.

---

[1] There are other methods, described in Montgomery and Peck (1992) and Myers (1990), that plot a modified version of the residual, called a partial residual, against each regressor. These partial residual plots are useful in displaying the relationship between the response y and each individual regressor.

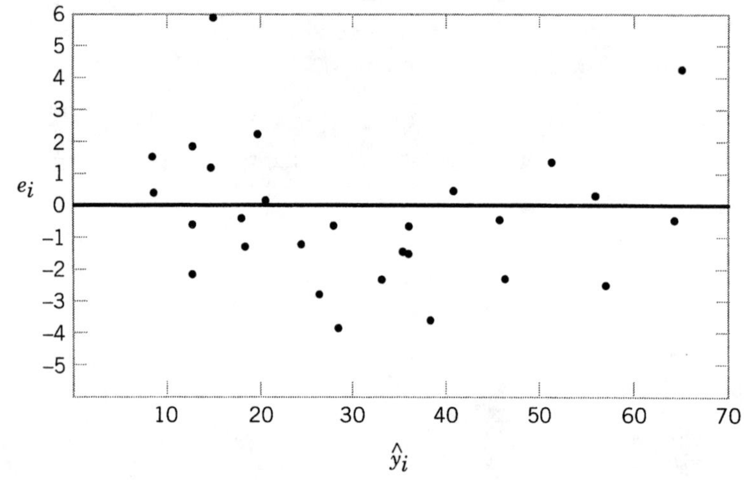

Figure 11-6 Plot of residuals against $\hat{y}$.

Many regression computer programs compute other types of scaled residuals. One of the most popular is the **studentized residual**

$$r_i = \frac{e_i}{\sqrt{\hat{\sigma}^2(1 - h_{ii})}} \qquad i = 1, 2, \ldots, n \qquad (11\text{-}40)$$

where h_{ii} is the ith diagonal element of the matrix

$$\mathbf{H} = \mathbf{X}(\mathbf{X}'\mathbf{X})^{-1}\mathbf{X}'$$

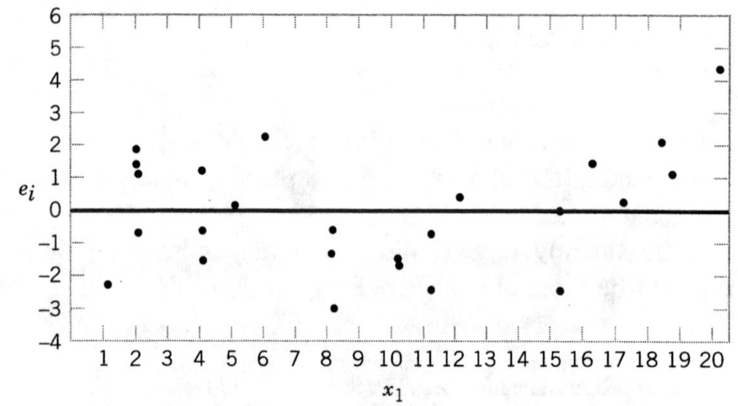

Figure 11-7 Plot of residuals against x_1.

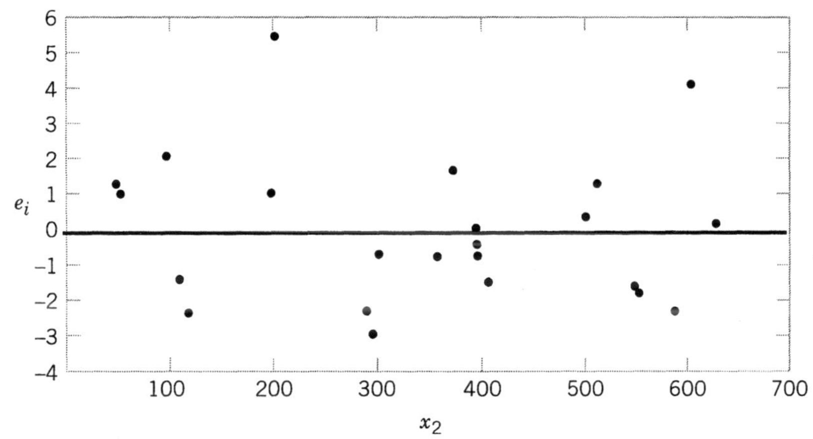

Figure 11-8 Plot of residuals against x_2.

The **H** matrix is sometimes called the "hat" matrix, since

$$\begin{aligned}\hat{\mathbf{y}} &= \mathbf{X}\hat{\boldsymbol{\beta}} \\ &= \mathbf{X}(\mathbf{X}'\mathbf{X})^{-1}\mathbf{X}'\mathbf{y} \\ &= \mathbf{H}\mathbf{y}\end{aligned}$$

Thus **H** transforms the observed values of **y** into a vector of fitted values $\hat{\mathbf{y}}$.

Since each row of the matrix **X** corresponds to a vector, say $\mathbf{x}_i' = [1, x_{i1}, x_{i2}, \ldots, x_{ik}]$, then another way to write the diagonal elements of the hat matrix is

$$h_{ii} = \mathbf{x}_i'(\mathbf{X}'\mathbf{X})^{-1}\mathbf{x}_i \tag{11-41}$$

Note that apart from σ^2, h_{ii} is the variance of the fitted value $\hat{y}_i$. The quantities h_{ii} were used in the computation of the confidence interval on the mean response in Section 11-2.

Under the usual assumptions that the model errors are independently distributed with mean zero and variance σ^2, we can show that the variance of the ith residual e_i is

$$V(e_i) = \sigma^2(1 - h_{ii}), \qquad i = 1, 2, \ldots, n$$

Furthermore, the h_{ii} elements must fall in the interval

$$0 < h_{ii} \leq 1$$

This implies that the standardized residuals understate the true residual magnitude; thus, the studentized residuals would be a better statistic to examine in evaluating potential outliers.

To illustrate, consider the two observations identified in Example 11-11 as having residuals that might be unusually large, observations 15 and 17. The standardized residuals are

$$d_{15} = \frac{e_{15}}{\sqrt{\hat{\sigma}^2}} = \frac{5.88}{\sqrt{5.2352}} = 2.57$$

and

$$d_{17} = \frac{e_{17}}{\sqrt{MS_E}} = \frac{4.33}{\sqrt{5.2352}} = 1.89$$

Now $h_{15,15} = 0.0737$ and $h_{17,17} = 0.2593$, so the studentized residuals are

$$r_{15} = \frac{e_{15}}{\sqrt{\hat{\sigma}^2(1 - h_{15,15})}} = \frac{5.88}{\sqrt{5.2352(1 - 0.0737)}} = 2.67$$

and

$$r_{17} = \frac{e_{17}}{\sqrt{\hat{\sigma}^2(1 - h_{17,17})}} = \frac{4.33}{\sqrt{5.2352(1 - 0.2593)}} = 2.20$$

Notice that the studentized residuals are larger than the corresponding standardized residuals. However, the studentized residuals are still not so large as to cause us serious concern about possible outliers.

11-8.3 Influential Observations

When using multiple regression, we occasionally find that some subset of the observations is unusually influential. Sometimes these influential observations are relatively far away from the vicinity where the rest of the data were collected. A hypothetical situation for two variables is depicted in Fig. 11-9, where one observation in x-space is remote from the rest of the data. The disposition of points in the x-space is important in determining the properties of the model. For example, point (x_{i1}, x_{i2}) in Fig. 11-9 may be very influential in determining R^2, the estimates of the regression coefficients, and the magnitude of the error mean square.

We would like to examine the influential points to determine whether they control many model properties. If these influential points are "bad" points, or erroneous in any way, then they should be eliminated. On the other hand, there may be nothing wrong with these points, but at least we would like to determine whether or not they produce results consistent with the rest of the data. In any event, even if an influential point is a valid one, if it controls important model properties, we would like to know this, since it could have an impact on the use of the model.

Montgomery and Peck (1992) and Myers (1990) describe several methods for detecting influential observations. An excellent diagnostic is the distance measure developed by

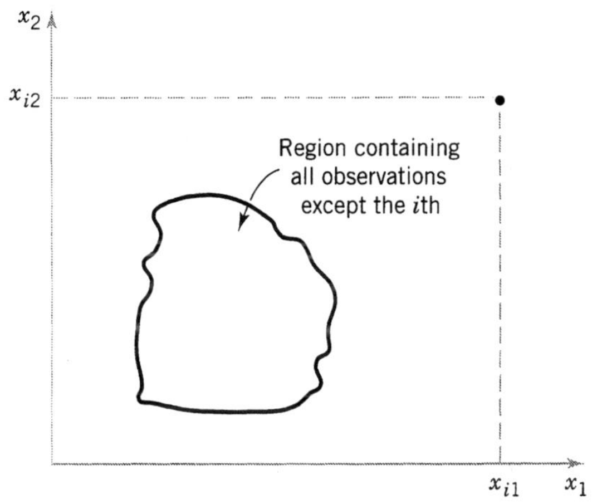

Figure 11-9 A point that is remote in x-space.

Dennis R. Cook. This is a measure of the squared distance between the usual least squares estimate of $\boldsymbol{\beta}$ based on all n observations and the estimate obtained when the ith point is removed, say, $\hat{\boldsymbol{\beta}}_{(i)}$. The **Cook distance measure** is

$$D_i = \frac{(\hat{\boldsymbol{\beta}}_{(i)} - \hat{\boldsymbol{\beta}})\mathbf{X}'\mathbf{X}(\hat{\boldsymbol{\beta}}_{(i)} - \hat{\boldsymbol{\beta}})}{p\hat{\sigma}^2} \qquad i = 1, 2, \ldots, n$$

Clearly, if the ith point is influential, its removal will result in $\hat{\boldsymbol{\beta}}_{(i)}$ changing considerably from the value $\hat{\boldsymbol{\beta}}$. Thus, a large value of D_i implies that the ith point is influential. The statistic D_i is actually computed using

$$D_i = \frac{r_i^2}{p}\frac{h_{ii}}{(1 - h_{ii})} \qquad i = 1, 2, \ldots, n \qquad (11\text{-}42)$$

From equation 11-42 we see that D_i consists of the squared studentized residual, which reflects how well the model fits the ith observation y_i [recall that $r_i = e_i/\sqrt{\hat{\sigma}^2(1 - h_{ii})}$] and a component that measures how far that point is from the rest of the data [$h_{ii}/(1 - h_{ii})$ is a measure of the distance of the ith point from the centroid of the remaining $n - 1$ points]. A value of $D_i > 1$ would indicate that the point is influential. Either component of D_i (or both) may contribute to a large value.

EXAMPLE 11-12

Table 11-7 lists the values of the hat matrix diagonals h_{ii} and Cook's distance measure D_i for the wire bond pull strength data in Example 11-1. To illustrate the calculations, consider the first observation:

$$\begin{aligned} D_1 &= \frac{r_1^2}{p} \cdot \frac{h_{11}}{(1 - h_{11})} \\ &= \frac{\left[e_1/\sqrt{MS_E(1 - h_{11})}\right]^2}{p} \cdot \frac{h_{11}}{(1 - h_{11})} \\ &= \frac{\left[1.57/\sqrt{5.2352(1 - 0.1573)}\right]^2}{3} \cdot \frac{0.1573}{(1 - 0.1573)} \\ &= 0.035 \end{aligned}$$

The Cook distance measure D_i does not identify any potentially influential observations in the data, for no value of D_i exceeds unity.

Table 11-7 Influence Diagnostics for the Wire Bond Pull Strength Data in Example 11-12

Observations i	h_{ii}	Cook's Distance Measure D_i
1	0.1573	0.035
2	0.1116	0.012
3	0.1419	0.060
4	0.1019	0.021
5	0.0418	0.024
6	0.0749	0.007
7	0.1181	0.036
8	0.1561	0.020
9	0.1280	0.160
10	0.0413	0.001
11	0.0925	0.013
12	0.0526	0.001
13	0.0820	0.001
14	0.1129	0.003
15	0.0737	0.187
16	0.0879	0.001
17	0.2593	0.565
18	0.2929	0.155
19	0.0962	0.018
20	0.1473	0.000
21	0.1296	0.052
22	0.1358	0.028
23	0.1824	0.002
24	0.1091	0.040
25	0.0729	0.000

EXERCISES FOR SECTION 11-8

11-36. Consider the regression model for the NFL data in Exercise 11-4.
(a) What proportion of total variability is explained by this model?
(b) Construct a normal probability plot of the residuals. What conclusion can you draw from this plot?
(c) Plot the residuals versus $\hat{y}$ and versus each regressor, and comment on model adequacy.
(d) Are there any influential points in these data?

11-37. Consider the gasoline mileage data in Exercise 11-5.
(a) What proportion of total variability is explained by this model?
(b) Construct a normal probability plot of the residuals and comment on the normality assumption.
(c) Plot residuals versus $\hat{y}$ and versus each regressor. Discuss these residual plots.
(d) Calculate Cook's distance for the observations in this data set. Are any observations influential?

11-38. Consider the electric power consumption data in Exercise 11-6.
(a) Calculate R^2 for this model. Interpret this quantity.
(b) Plot the residuals versus $\hat{y}$. Interpret this plot.
(c) Construct a normal probability plot of the residuals and comment on the normality assumption.

11-39. Consider the wear data in Exercise 11-7.
(a) Find the value of R^2 when the model uses the regressors x_1 and x_2.
(b) What happens to the value of R^2 when an interaction term x_1x_2 is added to the model? Does this necessarily imply that adding the interaction term is a good idea?

11-40. For the regression model for the wire bond pull strength data in Exercise 11-8.
(a) Plot the residuals versus $\hat{y}$ and versus the regressors used in the model. What information do these plots provide?
(b) Construct a normal probability plot of the residuals. Are there reasons to doubt the normality assumption for this model?
(c) Are there any indications of influential observations in the data?

11-41. Consider the semiconductor HFE data in Exercise 11-9.
(a) Plot the residuals from this model versus $\hat{y}$. Comment on the information in this plot.
(b) What is the value of R^2 for this model?
(c) Refit the model using log HFE as the response variable.
(d) Plot the residuals versus predicted log HFE for the model in part (c) above. Does this give any information about which model is preferable?
(e) Plot the residuals from the model in part (d) versus the regressor x_3. Comment on this plot.
(f) Refit the model to log HFE using x_1, x_2, and $1/x_3$, as the regressors. Comment on the effect of this change in the model.

11-42. Consider the regression model for the heat treating data in Exercise 11-10.
(a) Calculate the percent of variability explained by this model.
(b) Construct a normal probability plot for the residuals. Comment on the normality assumption.
(c) Plot the residuals versus $\hat{y}$ and interpret the display.
(d) Calculate Cook's distance for each observation and provide an interpretation of this statistic.

11-43. In Exercise 11-24 we fit a model to the response PITCH in the heat treating data of Exercise 11-10 using new regressors x_1 = SOAKTIME × SOAKPCT and x_2 = DIFFTIME × DIFFPCT.

(a) Calculate the R^2 for this model and compare it to the value of R^2 from the original model in Exercise 11-10. Does this provide some information about which model is preferable?

(b) Plot the residuals from this model versus $\hat{y}$ and on a normal probability scale. Comment on model adequacy.

(c) Find the values of Cook's distance measure. Are any observations unusually influential?

11-44. Consider the regression model for the NHL data from Exercise 11-11.

(a) Fit a model using "pts" as the only regressor.

(b) How much variability is explained by this model?

(c) Plot the residuals versus $\hat{y}$ and comment on model adequacy.

(d) Plot the residuals versus "PPG", the points scored while in power play. Does this indicate that the model would be better if this variable were included?

11-45. The diagonal elements of the hat matrix are often used to denote **leverage**—that is, a point that is unusual in its location in the x-space and that may be influential. Generally, the ith point is called a **leverage point** if its hat diagonal h_{ii} exceeds $2p/n$, which is twice the average size of all the hat diagonals. Recall that $p = k + 1$.

(a) Table 11-7 contains the hat diagonal for the wire bond pull strength data used in Example 11-1. Find the average size of these elements.

(b) Based on the criterion above, are there any observations that are leverage points in the data set?

11-9 POLYNOMIAL REGRESSION MODELS

The linear model $\mathbf{y} = \mathbf{X}\boldsymbol{\beta} + \boldsymbol{\epsilon}$ is a general model that can be used to fit any relationship that is **linear in the unknown parameters $\boldsymbol{\beta}$**. This includes the important class of **polynomial regression models.** For example, the second-degree polynomial in one variable

$$Y = \beta_0 + \beta_1 x + \beta_{11} x^2 + \epsilon \qquad (11\text{-}43)$$

and the second-degree polynomial in two variables

$$Y = \beta_0 + \beta_1 x_1 + \beta_2 x_2 + \beta_{11} x_1^2 + \beta_{22} x_2^2 + \beta_{12} x_1 x_2 + \epsilon \qquad (11\text{-}44)$$

are linear regression models.

Polynomial regression models are widely used when the response is curvilinear, because the general principles of multiple regression can be applied. The following example illustrates some of the types of analyses that can be performed.

EXAMPLE 11-13

Sidewall panels for the interior of an airplane are formed in a 1500-ton press. The unit manufacturing cost varies with the production lot size. The data shown below give the average cost per unit (in hundreds of dollars) for this product (y) and the production lot size (x). The scatter diagram, shown in Fig. 11-10, indicates that a second-order polynomial may be appropriate.

y	1.81	1.70	1.65	1.55	1.48	1.40	1.30	1.26	1.24	1.21	1.20	1.18
x	20	25	30	35	40	50	60	65	70	75	80	90

We will fit the model

$$Y = \beta_0 + \beta_1 x + \beta_{11} x^2 + \epsilon$$

The $\mathbf{y}$ vector, $\mathbf{X}$ matrix, and $\boldsymbol{\beta}$ vector are as follows:

$$\mathbf{y} = \begin{bmatrix} 1.81 \\ 1.70 \\ 1.65 \\ 1.55 \\ 1.48 \\ 1.40 \\ 1.30 \\ 1.26 \\ 1.24 \\ 1.21 \\ 1.20 \\ 1.18 \end{bmatrix} \qquad \mathbf{X} = \begin{bmatrix} 1 & 20 & 400 \\ 1 & 25 & 625 \\ 1 & 30 & 900 \\ 1 & 35 & 1225 \\ 1 & 40 & 1600 \\ 1 & 50 & 2500 \\ 1 & 60 & 3600 \\ 1 & 65 & 4225 \\ 1 & 70 & 4900 \\ 1 & 75 & 5625 \\ 1 & 80 & 6400 \\ 1 & 90 & 8100 \end{bmatrix} \qquad \boldsymbol{\beta} = \begin{bmatrix} \beta_0 \\ \beta_1 \\ \beta_{11} \end{bmatrix}$$

Solving the normal equations $\mathbf{X}'\mathbf{X}\hat{\boldsymbol{\beta}} = \mathbf{X}'\mathbf{y}$ gives the fitted model

$$\hat{y} = 2.19826629 - 0.02252236x + 0.00012507x^2$$

The test for significance of regression is shown in Table 11-8. Since $f_0 = 2171.07$ is significant at 1%, we conclude that at least one of the parameters β_1 and β_{11} is not zero. Furthermore, the standard tests for model adequacy do not reveal any unusual behavior, and we would conclude that this is a reasonable model for the sidewall panel cost data.

In fitting polynomials, we generally like to use the lowest-degree model consistent with the data. In this example, it would seem logical to investigate the possibility of dropping the quadratic term from the model. That is, we would like to test

$$H_0: \beta_{11} = 0$$

$$H_1: \beta_{11} \neq 0$$

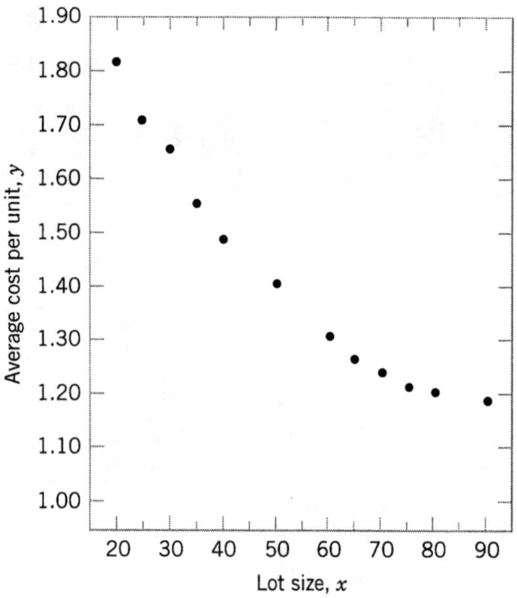

Figure 11-10 Data for Example 11-13.

The general regression significance test can be used to test this hypothesis. We need to determine the "extra sum of squares" due to β_{11}, or

$$SS_R(\beta_{11} \mid \beta_1,\beta_0) = SS_R(\beta_1,\beta_{11} \mid \beta_0) - SS_R(\beta_1 \mid \beta_0)$$

The sum of squares $SS_R(\beta_{11}, \beta_1 \mid \beta_0) = 0.5254$ from Table 11-9. To find $SS_R(\beta_1 \mid \beta_0)$, we fit a simple linear regression model to the original data, yielding

$$\hat{y} = 1.90036320 - 0.00910056x$$

It can be easily verified that the regression sum of squares for this model is

$$SS_R(\beta_1 \mid \beta_0) = 0.4942$$

Table 11-8 Test for Significance of Regression for the Second-Order Model in Example 11-13

Source of Variation	Sum of Squares	Degrees of Freedom	Mean Square	f_0	P-value
Regression	0.5254	2	0.262700	2171.07	5.18E-15
Error	0.0011	9	0.000121		
Total	0.5265	11			

Table 11-9 Analysis of Variance for Example 11-13, Showing the Test for H_0: $\beta_{11} = 0$

Source of Variation	Sum of Squares	Degrees of Freedom	Mean Square	f_0	P-value
Regression	$SS_R(\beta_1, \beta_{11} \mid \beta_0) = 0.5254$	2	0.262700	2171.07	5.18E-15
Linear	$SS_R(\beta_1 \mid \beta_0) = 0.4942$	1	0.494200	4084.30	1.17E-15
Quadratic	$SS_R(\beta_{11} \mid \beta_0, \beta_1) = 0.0312$	1	0.031200	258.18	5.51E-9
Error	0.0011	9	0.00121		
Total	0.5265	11			

Therefore, the extra sum of the squares due to β_{11}, given that β_1 and β_0 are in the model, is

$$
\begin{aligned}
SS_R(\beta_{11} \mid \beta_1,\beta_0) &= SS_R(\beta_1,\beta_{11} \mid \beta_0) - SS_R(\beta_1 \mid \beta_0) \\
&= 0.5254 - 0.4942 \\
&= 0.0312
\end{aligned}
$$

The analysis of variance, with the test of H_0: $\beta_{11} = 0$ incorporated into the procedure, is displayed in Table 11-9. Note that the quadratic term contributes significantly to the model.

11-10 INDICATOR VARIABLES

The regression models presented in previous sections have been based on **quantitative** variables, that is, variables that are measured on a numerical scale. For example, variables such as temperature, pressure, distance, and voltage are quantitative variables. Occasionally, we need to incorporate **qualitative** variables in a regression model. For example, suppose that one of the variables in a regression model is the operator who is associated with each observation y_i. Assume that only two operators are involved. We may wish to assign different levels to the two operators to account for the possibility that each operator may have a different effect on the response.

The usual method of accounting for the different levels of a qualitative variable is to use **indicator variables.** For example, to introduce the effect of two different operators into a regression model, we could define an indicator variable as follows:

$$
x = \begin{cases} 0 \text{ if the observation is from operator 1} \\ 1 \text{ if the observation is from operator 2} \end{cases}
$$

In general, a qualitative variable with t-levels can be modeled by $t - 1$ indicator variables, which are assigned the value of either zero or one. Thus, if there are *three* operators, the different levels will be accounted for by the *two* indicator variables defined as follows:

x_1	x_2	
0	0	if the observation is from operator 1
1	0	if the observation is from operator 2
0	1	if the observation is from operator 3

Indicator variables are also referred to as **dummy** variables. The following example (from Montgomery and Peck, 1992) illustrates some of the uses of indicator variables; for other applications, see Montgomery and Peck (1992).

EXAMPLE 11-14

A mechanical engineer is investigating the surface finish of metal parts produced on a lathe and its relationship to the speed (in RPM) of the lathe. The data are shown in Table 11-10. Note that the data have been collected using two different types of cutting tools. Since the type of cutting tool likely affects the surface finish, we will fit the model

$$Y = \beta_0 + \beta_1 x_1 + \beta_2 x_2 + \epsilon$$

Table 11-10 Surface Finish Data for Example 11-14

Observation Number, i	Surface Finish y_i	RPM	Type of Cutting Tool
1	45.44	225	302
2	42.03	200	302
3	50.10	250	302
4	48.75	245	302
5	47.92	235	302
6	47.79	237	302
7	52.26	265	302
8	50.52	259	302
9	45.58	221	302
10	44.78	218	302
11	33.50	224	416
12	31.23	212	416
13	37.52	248	416
14	37.13	260	416
15	34.70	243	416
16	33.92	238	416
17	32.13	224	416
18	35.47	251	416
19	33.49	232	416
20	32.29	216	416

where Y is the surface finish, x_1 is the lathe speed in RPM, and x_2 is an indicator variable denoting the type of cutting tool used; that is,

$$x_2 = \begin{cases} 0, \text{ for tool type } 302 \\ 1, \text{ for tool type } 416 \end{cases}$$

The parameters in this model may be easily interpreted. If $x_2 = 0$, then the model becomes

$$Y = \beta_0 + \beta_1 x_1 + \epsilon$$

which is a straight-line model with slope β_1 and intercept β_0. However, if $x_2 = 1$, then the model becomes

$$Y = \beta_0 + \beta_1 x_1 + \beta_2(1) + \epsilon = \beta_0 + \beta_2 + \beta_1 x_1 + \epsilon$$

which is a straight-line model with slope β_1 and intercept $\beta_0 + \beta_2$. Thus, the model $Y = \beta_0 + \beta_1 x + \beta_2 x_2 + \epsilon$ implies that surface finish is linearly related to lathe speed and that the slope β_1 does not depend on the type of cutting tool used. However, the type of cutting tool does affect the intercept, and β_2 indicates the change in the intercept associated with a change in tool type from 302 to 416.

The **X** matrix and **y** vector for this problem are as follows:

$$\mathbf{X} = \begin{bmatrix} 1 & 225 & 0 \\ 1 & 200 & 0 \\ 1 & 250 & 0 \\ 1 & 245 & 0 \\ 1 & 235 & 0 \\ 1 & 237 & 0 \\ 1 & 265 & 0 \\ 1 & 259 & 0 \\ 1 & 221 & 0 \\ 1 & 218 & 0 \\ 1 & 224 & 1 \\ 1 & 212 & 1 \\ 1 & 248 & 1 \\ 1 & 260 & 1 \\ 1 & 243 & 1 \\ 1 & 238 & 1 \\ 1 & 224 & 1 \\ 1 & 251 & 1 \\ 1 & 232 & 1 \\ 1 & 216 & 1 \end{bmatrix} \qquad \mathbf{y} = \begin{bmatrix} 45.44 \\ 42.03 \\ 50.10 \\ 48.75 \\ 47.92 \\ 47.79 \\ 52.26 \\ 50.52 \\ 45.58 \\ 44.78 \\ 33.50 \\ 31.23 \\ 37.52 \\ 37.13 \\ 34.70 \\ 33.92 \\ 32.13 \\ 35.47 \\ 33.49 \\ 32.29 \end{bmatrix}$$

The fitted model is

$$\hat{y} = 14.27620 + 0.14115x_1 - 13.28020x_2$$

The analysis of variance for this model is shown in Table 11-11. Note that the hypothesis H_0: $\beta_1 = \beta_2 = 0$ (significance of regression) would be rejected at any reasonable level of significance because the P-value is very small. This table also contains the sums of squares

$$\begin{aligned} SS_R &= SS_R(\beta_1, \beta_2 \mid \beta_0) \\ &= SS_R(\beta_1 \mid \beta_0) + SS_R(\beta_2 \mid \beta_1, \beta_0) \end{aligned}$$

so that a test of the hypothesis H_0: $\beta_2 = 0$ can be made. Since this hypothesis is also rejected, we conclude that tool type has an effect on surface finish.

It is also possible to use indicator variables to investigate whether tool type affects both the slope and intercept. Let the model be

$$Y = \beta_0 + \beta_1 x_1 + \beta_2 x_2 + \beta_3 x_1 x_2 + \epsilon$$

where x_2 is the indicator variable. Now if tool type 302 is used, $x_2 = 0$, and the model is

$$Y = \beta_0 + \beta_1 x_1 + \epsilon$$

If tool type 416 is used, $x_2 = 1$, and the model becomes

$$\begin{aligned} Y &= \beta_0 + \beta_1 x_1 + \beta_2 + \beta_3 x_1 + \epsilon \\ &= (\beta_0 + \beta_2) + (\beta_1 + \beta_3)x_1 + \epsilon \end{aligned}$$

Note that β_2 is the change in the intercept and that β_3 is the change in slope produced by a change in tool type.

Table 11-11 Analysis of Variance of Example 11-14

Source of Variation	Sum of Squares	Degrees of Freedom	Mean Square	f_0	P-value
Regression	1012.0595	2	506.0297	1103.69	1.02 E-18
$SS_R(\beta_1 \mid \beta_0)$	130.6091	1	130.6091	284.87	4.70 E-12
$SS_R(\beta_2 \mid \beta_1, \beta_0)$	881.4504	1	881.4504	1922.52	6.24 E-19
Error	7.7943	17	0.4508		
Total	1019.8538	19			

Another method of analyzing these data is to fit separate regression models to the data for each tool type. However, the indicator variable approach has several advantages. First, only one regression model must be fit. Second, by pooling the data on both tool types, more degrees of freedom for error are obtained. Third, tests of both hypotheses on the parameters β_2 and β_3 are just special cases of the general regression significance test.

EXERCISES FOR SECTIONS 11-9 AND 11-10

11-46. An article entitled "A Method for Improving the Accuracy of Polynomial Regression Analysis" in the *Journal of Quality Technology* (1971, pp. 149–155) reported the following data on y = ultimate shear strength of a rubber compound (psi) and x = cure temperature (°F).

y	770	800	840	810
x	280	284	292	295

y	735	640	590	560
x	298	305	308	315

(a) Fit a second-order polynomial to these data.
(b) Test for significance of regression using $\alpha = 0.05$.
(c) Test the hypothesis that $\beta_{11} = 0$ using $\alpha = 0.05$.
(d) Compute the residuals from part (a) and use them to evaluate model adequacy.

11-47. Consider the following data, which result from an experiment to determine the effect of x = test time in hours at a particular temperature on y = change in oil viscosity.

y	−1.42	−1.39	−1.55	−1.89	−2.43
x	.25	.50	.75	1.00	1.25

y	−3.15	−4.05	−5.15	−6.43	−7.89
x	1.50	1.75	2.00	2.25	2.50

(a) Fit a second-order polynomial to the data.
(b) Test for significance of regression using $\alpha = 0.05$.
(c) Test the hypothesis that $\beta_{11} = 0$ using $\alpha = 0.05$.
(d) Compute the residuals from part (a) and use them to evaluate model adequacy.

11-48. When fitting polynomial regression models, we often subtract $\bar{x}$ from each x value to produce a "centered" regressor $x' = x - \bar{x}$. This reduces the effects of dependencies among the model terms and often leads to more accurate estimates of the regression coefficients. Using the data from Exercise 11-46, fit the model $Y = \beta_0^* + \beta_1^* x' + \beta_{11}^* (x')^2 + \epsilon$. Use the results to estimate the coefficients in the uncentered model $Y = \beta_0 + \beta_1 x + \beta_{11} x^2 + \epsilon$.

11-49. Suppose that we use a standardized variable $x' = (x - \bar{x})/s_x$, where s_x is the standard deviation of x, in constructing a polynomial regression model. Using the data in Exercise 11-46 and the standardized variable approach, fit the model $Y = \beta_0^* + \beta_1^* x' + \beta_{11}^* (x')^2 + \epsilon$.

(a) What value of y do you predict when $x = 285$°F?
(b) Estimate the regression coefficients in the unstandardized model $Y = \beta_0 + \beta_1 x + \beta_{11} x^2 + \epsilon$.
(c) What can you say about the relationship between SS_E and R^2 for the standardized and unstandardized models?
(d) Suppose that $y' = (y - \bar{y})/s_y$ is used in the model along with x'. Fit the model and comment on the relationship between SS_E and R^2 in the standardized model and the unstandardized model.

11-50. The data shown below were collected during an experiment to determine the change in thrust efficiency (y, in percent) as the divergence angle of a rocket nozzle (x) changes.

y	24.60	24.71	23.90	39.50	39.60	57.12
x	4.0	4.0	4.0	5.0	5.0	6.0

y	67.11	67.24	67.15	77.87	80.11	84.67
x	6.5	6.5	6.75	7.0	7.1	7.3

(a) Fit a second-order model to the data.
(b) Test for significance of regression and lack of fit using $\alpha = 0.05$.
(c) Test the hypothesis that $\beta_{11} = 0$, using $\alpha = 0.05$.
(d) Plot the residuals and comment on model adequacy.
(e) Fit a cubic model, and test for the significance of the cubic term using $\alpha = 0.05$.

11-51. An article in the *Journal of Pharmaceuticals Sciences* (Vol. 80, 1991, pp. 971–977) presents data on the observed mole fraction solubility of a solute at a constant temperature and the dispersion, dipolar, and hydrogen bonding Hansen partial solubility parameters. The data are as shown in the following table, where y is the negative logarithm of the mole fraction solubility, x_1 is the dispersion partial solubility, x_2 is the dipolar partial solubility, and x_3 is the hydrogen bonding partial solubility.

(a) Fit the model $Y = \beta_0 + \beta_1 x_1 + \beta_2 x_2 + \beta_3 x_3 + \beta_{12} x_1 x_2 + \beta_{13} x_1 x_3 + \beta_{23} x_2 x_3 + \beta_{11} x_1^2 + \beta_{22} x_2^2 + \beta_{33} x_3^2 + \epsilon$.
(b) Test for significance of regression using $\alpha = 0.05$.
(c) Plot the residuals and comment on model adequacy.
(d) Use the extra sum of squares method to test the contribution of the second-order terms using $\alpha = 0.05$.

Observation Number	y	x_1	x_2	x_3
1	0.22200	7.3	0.0	0.0
2	0.39500	8.7	0.0	0.3
3	0.42200	8.8	0.7	1.0
4	0.43700	8.1	4.0	0.2
5	0.42800	9.0	0.5	1.0
6	0.46700	8.7	1.5	2.8
7	0.44400	9.3	2.1	1.0
8	0.37800	7.6	5.1	3.4
9	0.49400	10.0	0.0	0.3
10	0.45600	8.4	3.7	4.1
11	0.45200	9.3	3.6	2.0
12	0.11200	7.7	2.8	7.1
13	0.43200	9.8	4.2	2.0
14	0.10100	7.3	2.5	6.8
15	0.23200	8.5	2.0	6.6
16	0.30600	9.5	2.5	5.0
17	0.09230	7.4	2.8	7.8
18	0.11600	7.8	2.8	7.7
19	0.07640	7.7	3.0	8.0
20	0.43900	10.3	1.7	4.2
21	0.09440	7.8	3.3	8.5
22	0.11700	7.1	3.9	6.6
23	0.07260	7.7	4.3	9.5
24	0.04120	7.4	6.0	10.9
25	0.25100	7.3	2.0	5.2
26	0.00002	7.6	7.8	20.7

11-52. Consider the gasoline mileage data in Exercise 11-5.

(a) Discuss how you would model the information about the type of transmission in the car.
(b) Fit a regression model to the gasoline mileage using engine displacement, horsepower, and the type of transmission in the car as the regressors.
(c) Is there evidence that the type of transmission affects gasoline mileage performance?

11-53. Consider the tool life data in Example 11-14. Test the hypothesis that two different regression models (with different slopes and intercepts) are required to adequately model the data. Use indicator variables in answering this question.

11-11 SELECTION OF VARIABLES IN MULTIPLE REGRESSION

11-11.1 Model-Building Problem

An important problem in many applications of regression analysis involves selecting the set of independent or regressor variables to be used in the model. Sometimes previous experience or underlying theoretical considerations can help the analyst specify the set of regressor variables he or she should use in a particular situation. Usually, however, the problem consists of selecting an appropriate set of regressors from a set that quite likely includes all the important variables, but we are sure that not all these candidate regressors are necessary to adequately model the response Y.

In such a situation, we are interested in screening the candidate variables to obtain a regression model that contains the "best" subset of regressor variables. We would like the final model to contain enough regressor variables so that in the intended use of the model (prediction, for example) it will perform satisfactorily. On the other hand, to keep model maintenance costs to a minimum and to make the model easy to use, we would like the model to use as few regressor variables as possible. The compromise between these conflicting objectives is often called finding the "best" regression equation. However, in most problems, no single regression model is "best" in terms of the various evaluation criteria that have been proposed. A great deal of judgment and experience with the system being modeled is usually necessary to select an appropriate set of regressor variables for a regression equation.

No single algorithm will always produce a good solution to the variable selection problem. Most of the currently available procedures are search techniques, and to perform satisfactorily, they require interaction with and judgment by the analyst. We now briefly discuss some of the more popular variable selection techniques.

11-11.2 Computational Procedures for Variable Selection

We assume that there are K candidate regressors, $x_1, x_2, \ldots, x_K$, and a single response variable y. All models will include an intercept term β_0, so that the model with *all* variables included would have $K + 1$ terms. Furthermore, the functional form of each candidate variable (for example, $x_1 = 1/x$, $x_2 = \ln x$, etc.) is correct.

All Possible Regressions

This approach requires that the analyst fit all the regression equations involving one candidate variable, all regression equations involving two candidate variables, and so on. Then these equations are evaluated according to some suitable criteria to select the "best" regression model. If there are K candidate regressors, there are 2^K total equations to be examined. For example, if $K = 4$, there are $2^4 = 16$ possible regression equations; while if $K = 10$, there are $2^{10} = 1024$ possible regression equations. Hence, the number of equations to be examined increases rapidly as the number of candidate variables increases.

A number of criteria may be used for evaluating and comparing the different regression models obtained. A commonly used criterion is based on the coefficient of multiple determination. Let R_p^2 denote the coefficient of determination for a regression model with p terms, that is, $p - 1$ regressors and an intercept term (note that $p \leq K + 1$). Computationally, we have

$$R_p^2 = \frac{SS_R(p)}{SS_T} = 1 - \frac{SS_E(p)}{SS_T} \tag{11-45}$$

where $SS_R(p)$ and $SS_E(p)$ denote the regression sum of squares and the error sum of squares, respectively, for a p-variable model. Now R_p^2 increases as p increases and is a maximum when $p = K + 1$. Therefore, the analyst adds variables to the model up to the point where an additional variable is not useful because it results in a small increase in R_p^2. The general approach is illustrated in Fig. 11-11, which gives a hypothetical plot of R_p^2 against p. Typically, the analyst examines a display such as this and chooses the number of regressors in the model as the point at which the "knee" in the curve becomes apparent. Clearly, making this choice requires judgment on the part of the analyst.

A second criterion is to consider the mean square error for a p-variable equation, denoted as $MS_E(p) = SS_E(p)/(n - p)$. Generally, $MS_E(p)$ decreases as p increases, but this is not necessarily so. If the addition of a regressor to the model with $p - 1$ terms does not reduce the error sum of squares by an amount equal to the error mean square in the old $p - 1$ term model, $MS_E(p)$ will *increase* because of the loss of one degree of freedom for error. Therefore, a logical criterion is to choose the regressors so that $MS_E(p)$ is a minimum. Alternatively, since $MS_E(p)$ is usually relatively flat in the vicinity of the minimum, we could choose a model so that adding more regressors to the model would produce only very small reductions in $MS_E(p)$. The general procedure is illustrated in Fig. 11-12.

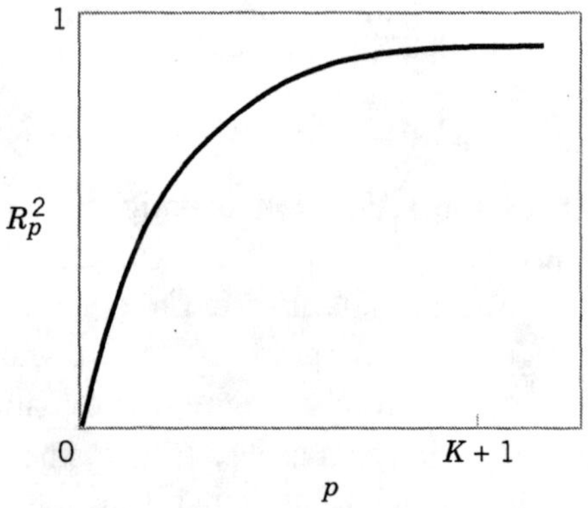

Figure 11-11 Plot of R_p^2 against p.

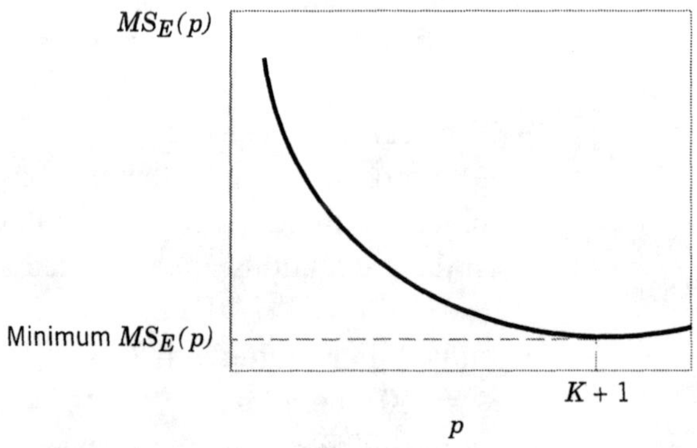

Figure 11-12 Plot of MS_E (p) against p.

A third criterion is the C_p statistic, which is a measure of the total mean square error for the regression model. We define the total standardized mean square error for the regression model as

$$\begin{aligned}\Gamma_p &= \frac{1}{\sigma^2}\sum_{i=1}^{n} E[\hat{Y}_i - E(Y_i)]^2 \\ &= \frac{1}{\sigma^2}\left\{\sum_{i=1}^{n}[E(Y_i) - E(\hat{Y}_i)]^2 + \sum_{i=1}^{n} V(\hat{Y}_i)\right\} \\ &= \frac{1}{\sigma^2}[(\text{bias})^2 + \text{variance}]\end{aligned}$$

We use the mean square error from the *full* $K + 1$ term model as an estimate of σ^2; that is, $\hat{\sigma}^2 = MS_E(K + 1)$. Then an estimator of Γ_p is [see Montgomery and Peck (1992) or Myers (1990) for the details]

$$C_p = \frac{SS_E(p)}{\hat{\sigma}^2} - n + 2p \qquad (11\text{-}46)$$

If the p-term model has negligible bias, then it can be shown that

$$E(C_p \mid \text{zero bias}) = p$$

Therefore, the values of C_p for each regression model under consideration should be evaluated relative to p. The regression equations that have negligible bias will have values of C_p that are close to p, while those with significant bias will have values of C_p that are significantly greater than p. We then choose as the "best" regression equation either a model with *minimum* C_p or a model with a slightly larger C_p that does not contain as much bias (i.e., $C_p \cong p$).

Another criterion is based on a modification of R_p^2 that accounts for the number of variables in the model. This statistic is called the **adjusted** R_p^2 defined as

$$\overline{R}_p^2 = 1 - \frac{n-1}{n-p}(1 - R_p^2) \qquad (11\text{-}47)$$

Note that $\overline{R}_p^2$ may decrease as p increases if the decrease in $(n - 1)(1 - \overline{R}_p^2)$ is not compensated for by the loss of one degree of freedom in $n - p$. The experimenter would usually select the regression model that has the maximum value of $\overline{R}_p^2$. However, note that this is equivalent to the model that minimizes $MS_E(p)$, since

$$\begin{aligned}\overline{R}_p^2 &= 1 - \left(\frac{n-1}{n-p}\right)(1 - R_p^2) \\ &= 1 - \left(\frac{n-1}{n-p}\right)\frac{SS_E(p)}{SS_T} \\ &= 1 - \left(\frac{n-1}{SS_T}\right) MS_E(p)\end{aligned}$$

EXAMPLE 11-15

The data in Table 11-12 are an expanded set of data for the wire bond pull strength study in Example 11-1. There are now four candidate variables: wire length (x_1), die height (x_2), post height (x_3), loop length (x_4).

Table 11-13 presents the results of running all possible regressions (except the trivial model with only an intercept) on these data. The values of R_p^2, $\overline{R}_p^2$, $SS_R(p)$, $SS_E(p)$,

Table 11-12 Wire Bond Pull Strength Data for Example 11-15

Observation Number	Pull Strength y	Wire Length x_1	Die Height x_2	Post Height x_3	Loop Length x_4
1	9.95	2	50	1	1
2	24.45	8	110	1	1
3	31.75	11	120	2	1
4	35.00	10	550	2	2
5	25.02	8	295	1	1
6	16.86	4	200	1	1
7	14.38	2	375	1	1
8	9.60	2	52	1	1
9	24.35	9	100	1	1
10	27.50	8	300	2	1
11	17.08	4	412	2	2
12	37.00	11	400	3	2
13	41.95	12	500	3	3
14	11.66	2	360	1	1
15	21.65	4	205	2	2
16	17.89	4	400	2	1
17	69.00	20	600	4	4
18	10.30	1	585	1	1
19	34.93	10	540	2	1
20	46.59	15	250	3	2
21	44.88	15	290	3	1
22	54.12	16	510	3	3
23	56.63	17	590	2	2
24	22.13	6	100	2	1
25	21.15	5	400	1	1

Table 11-13 All Possible Regressions for the Data in Example 11-15

Number of Variables in Model	p	Variables in Model	R_p^2	$SS_R(p)$	$SS_E(p)$	$MS_E(p)$	$\overline{R}_p^2$	C_p
1	2	x_2	.24291747	1483.2406	4622.7041	200.9871	.2100	1479.93
1	2	x_4	.56007492	3419.7865	2686.1582	116.7895	.5409	851.16
1	2	x_3	.39830969	4263.8404	1842.1043	80.0915	.6852	577.11
1	2	x_1	.96395437	5885.8521	220.0926	9.5692	.9624	50.46
2	3	x_2, x_4	.57158123	3490.0434	2615.9013	118.9046	.5326	830.35
2	3	x_2, x_3	.72162460	4406.1999	1699.7448	77.2611	.6963	532.88
2	3	x_3, x_4	.72232683	4410.4877	1695.4570	77.0662	.6971	531.49
2	3	x_1, x_3	.97220237	5936.2139	169.7308	7.7150	.9697	36.11
2	3	x_1, x_2	.98113748	5990.7712	115.1735	5.2352	.9794	18.40
2	3	x_1, x_4	.98302963	6002.3246	103.6201	4.7100	.9815	14.64
3	4	x_2, x_3, x_4	.73299076	4475.6010	1630.3437	77.6354	.6948	512.35
3	4	x_1, x_3, x_4	.98327895	6003.8469	102.0978	4.8618	.9809	16.15
3	4	x_1, x_2, x_3	.98517507	6015.4245	90.5203	4.3105	.9831	12.39
3	4	x_1, x_2, x_4	.98961493	6042.5340	63.4107	3.0196	.9881	3.59
4	5	x_1, x_2, x_3, x_4	.98991196	6044.3477	61.5970	3.0799	.9879	5

$MS_E(p)$, and C_p are also shown in this table, and a plot of R_p^2 against p is shown in Fig. 11-13. In terms of R^2 improvement, little is gained in going from a two-variable to a three-variable model. A plot of the minimum $MS_E(p)$ for each subset of size p is shown in Fig. 11-14. Several models have good values of $MS_E(p)$. The best two-variable model is either (x_1, x_2) or (x_1, x_4), and the best three-variable model is (x_1, x_2, x_4). The minimum value of $MS_E(p)$ occurs for the three-variable model (x_1, x_2, x_4). Although several other models have relatively small values of $MS_E(p)$, such as (x_1, x_2, x_3) and (x_1, x_2), the model with regressors (x_1, x_2, x_4) is definitely superior with respect to the $MS_E(p)$ criterion. Note that, as expected, this model also minimizes the adjusted R_p^2. A plot of the C_p statistic is shown in Fig. 11-15. The only model with $C_p \leq p$ is the three-variable equation with regressors (x_1, x_2, x_4). To illustrate the calculations, for this equation we would find that

$$C_p = \frac{SS_E(p)}{\hat{\sigma}^2} - n + 2p$$

$$= \frac{63.4107}{3.0799} - 25 + 2(4) = 3.59$$

noting that $\hat{\sigma}^2 = 3.0799$ is obtained from the *full* equation (x_1, x_2, x_3, x_4). Since all other models contain substantial bias $(C_p > p)$, we would conclude on the basis of the C_p criterion that the best subset of the regressor variables is (x_1, x_2, x_4). Since this model also results in a minimum $MS_E\ (p)$ and a high R_p^2, we would select it as the "best" regression equation. The final model is

$$\hat{y} = 1.36707 + 2.53492x_1 + 0.0085x_2 + 2.59928x_4$$

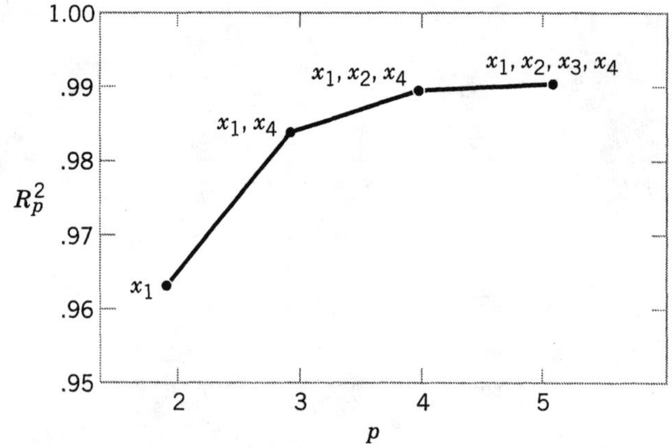

Figure 11-13 The R_p^2 plot for Example 11-15.

The all-possible regression approach requires considerable computational effort, even when K is moderately small. However, if the analyst is willing to look at something less than the fitted model coefficients and all of the associated model statistics, it is possible to devise algorithms for all possible regressions that produce less information about each model but that are more efficient computationally. For example, suppose that we could efficiently calculate only the MS_E for each model. Since the models with large MS_E are not likely to be selected as the best regression equations, we would only then have to examine in detail the models with small values of MS_E. Several approaches can be used to develop a computationally efficient algorithm for all possible regressions. For more discussion of these algorithms and illustrations of their use, see Montgomery and Peck (1992) and Myers (1990). Typically, problems with up to about 30 candidate variables can be easily solved by this approach.

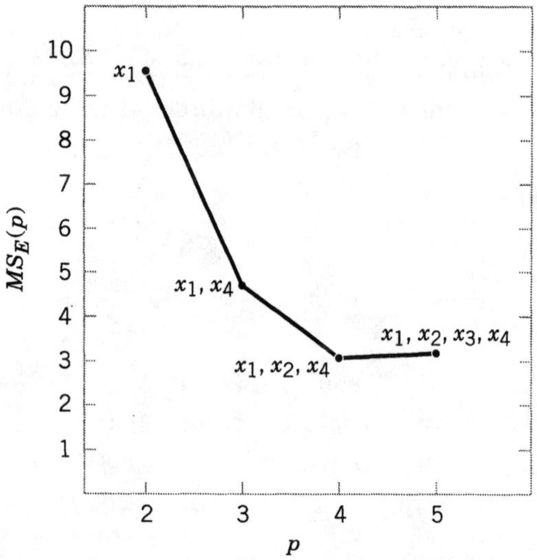

Figure 11-14 The MS_E (p) plot for Example 11-15.

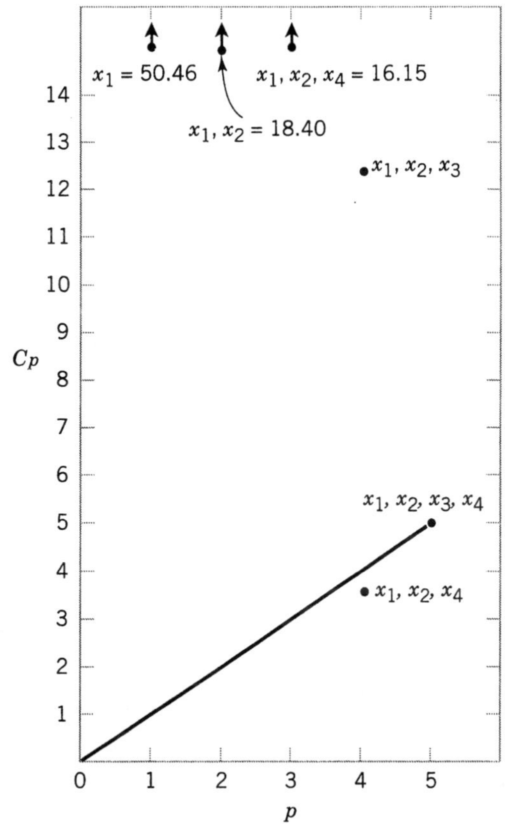

Figure 11-15 The C_p plot for Example 11-15.

Stepwise Regression

Stepwise regression is probably the most widely used variable selection technique. The procedure iteratively constructs a sequence of regression models by adding or removing variables at each step. The criterion for adding or removing a variable at any step is usually expressed in terms of a partial F-test. Let f_{in} be the value of the F-random variable for adding a variable to the model, and let f_{out} be the value of the F-random variable for removing a variable from the model. We must have $f_{in} \geq f_{out}$, and usually $f_{in} = f_{out}$.

Stepwise regression begins by forming a one-variable model using the regressor variable that has the highest correlation with the response variable Y. This will also be the regressor producing the largest F-statistic. For example, suppose that at this step, x_1 is selected. At the second step, the remaining $K - 1$ candidate variables are examined, and the variable for which the partial F-statistic

$$F_j = \frac{SS_R(\beta_j \mid \beta_1, \beta_0)}{MS_E(x_j, x_1)} \tag{11-48}$$

is a maximum is added to the equation, provided that $f_j > f_{in}$. In equation 11-48, $MS_E(x_j, x_1)$ denotes the mean square for error for the model containing both x_1 and x_j. Suppose that this procedure indicates that x_2 should be added to the model. Now the stepwise regression algorithm determines whether the variable x_1 added at the first step should be removed. This is done by calculating the F-statistic

$$F_1 = \frac{SS_R(\beta_1 \mid \beta_2, \beta_0)}{MS_E(x_1, x_2)} \tag{11-49}$$

If the calculated value $f_1 < f_{\text{out}}$, the variable x_1 is removed; otherwise it is retained, and we would attempt to add a regressor to the model containing both x_1 and x_2.

In general, at each step the set of remaining candidate regressors is examined, and the regressor with the largest partial F-statistic is entered, provided that the observed value of f exceeds f_{in}. Then the partial F-statistic for each regressor in the model is calculated, and the regressor with the smallest observed value of F is deleted if the observed $f < f_{\text{out}}$. The procedure continues until no other regressors can be added to or removed from the model.

Stepwise regression is almost always performed using a computer program. The analyst exercises control over the procedure by the choice of f_{in} and f_{out}. Some stepwise regression computer programs require that numerical values be specified for f_{in} and f_{out}. Since the number of degrees of freedom on MS_E depends on the number of variables in the model, which changes from step to step, a fixed value of f_{in} and f_{out} causes the type I and type II error rates to vary. Some computer programs allow the analyst to specify the type I error levels for f_{in} and f_{out}. However, the "advertised" significance level is not the true level, because the variable selected is the one that maximizes (or minimizes) the partial F-statistic at that stage. Sometimes it is useful to experiment with different values of f_{in} and f_{out} (or different advertised type I error rates) in several different runs to see if this substantially affects the choice of the final model.

EXAMPLE 11-16

We will apply stepwise regression to the wire bond pull strength data in Table 11-12. You may wish to refer to the computer output in Section 11-11.3, which supports the computations. Instead of specifying numerical values of f_{in} and f_{out}, we use an advertised type I error of $\alpha = 0.10$. The first step consists of building a simple linear regression model using the variable that gives the largest F-statistic. This is x_1, and since

$$f_1 = \frac{SS_R(\beta_1 \mid \beta_0)}{MS_E(x_1)} = \frac{5885.8521}{9.5692} = 615.08 > f_{\text{in}} = f_{0.10,1,23} = 2.94$$

x_1 is entered into the model.

The second step begins by finding the variable x_j that has the largest partial F-statistic, given that x_1 is in the model. This is x_4, and since

$$f_4 = \frac{SS_R(\beta_4 \mid \beta_1, \beta_0)}{MS_E(x_4, x_1)} = \frac{116.4725}{4.7100} = 24.73 > f_{\text{in}} = f_{0.10,1,22} = 2.95$$

x_4 is added to the model. Now the procedure evaluates whether or not x_1 should be retained, given that x_4 is in the model. This involves calculating

$$f_1 = \frac{SS_R(\beta_1 \mid \beta_4, \beta_0)}{MS_E(x_4, x_1)} = \frac{2582.5381}{4.7100} = 548.31 > f_{\text{out}} = f_{0.10,1,22} = 2.95$$

Therefore, x_1 should be retained. Step 2 terminates with both x_1 and x_4 in the model.

The third step finds the next variable for entry as x_2. Since

$$f_2 = \frac{SS_R(\beta_2 \mid \beta_1, \beta_4, \beta_0)}{MS_E(x_2, x_4, x_1)} = \frac{40.2094}{3.0196} = 13.32 > f_{\text{in}} = f_{0.10,1,21} = 2.96$$

x_2 is added to the model. Partial F-tests on x_1 (given x_2 and x_4) and x_4 (given x_2 and x_1) indicate that these regressors should be retained. Therefore, the third step concludes with the variables x_1, x_2, and x_4 in the model.

At the fourth step, the procedure attempts to add x_3. However, the partial F-statistic for x_3 is $f_3 = 0.59$, which is less than $f_{\text{in}} = f_{0.10,1,20} = 2.97$. Therefore, since this variable cannot be added and since there are no other candidate regressors to consider, the procedure terminates.

The stepwise regression procedure would conclude that the three-variable model with x_1, x_2, and x_4 is the best regression equation. However, the standard checks of model adequacy, such as residual analysis and evaluation of C_p plots, should be applied to the equation to ensure that the final model is a reasonable one, and that none of the regression assumptions are seriously violated. Note that this is the same regression model found by the all-possible regression method.

Forward Selection

The forward selection procedure is a variation of stepwise regression and is based on the principle that regressors should be added to the model one at a time until there are no remaining candidate regressors that produce a significant increase in the regression sum of squares. That is, variables are added one at a time as long as their partial F-value exceeds f_{in}. Forward selection is a simplification of stepwise regression that omits the partial F-test for deleting variables from the model that have been added at previous steps. This is a potential weakness of forward selection; that is, the procedure does not explore the effect that adding a regressor at the current step has on regressor variables added at earlier steps.

EXAMPLE 11-17

Application of the forward selection algorithm to the wire bond pull strength data in Table 11-12 would begin by adding x_1 to the model. Then, the regressor that induces the largest partial F-test, given that x_1 is in the model, is added—this is regressor x_4. The third step enters x_2, which produces the largest partial F-statistic, given that x_1 and x_4 are in the model. Since the partial F-statistic for x_3 is not significant, the procedure terminates. Note that in this example forward selection leads to the same final model as stepwise regression. This will not always happen. Sometimes forward selection produces a different model than does stepwise regression.

Backward Elimination

This algorithm begins with all K candidate regressors in the model. Then the regressor with the smallest partial F-statistic is deleted, if this F-statistic is insignificant; that is, if $f < f_{\text{out}}$. Next, the model with $K - 1$ regressors is fit, and the next regressor for potential elimination is found. The algorithm terminates when no further regressor can be deleted.

EXAMPLE 11-18

To apply backward elimination to the data in Table 11-12, we begin by fitting the full model in all four regressors. This model is

$$\hat{y} = 1.06813 + 2.48423x_1 + 0.00855x_2 + 0.62337x_3 + 2.29392x_4$$

The partial F-tests for each variable are

$$f_1 = \frac{SS_R(\beta_1 \mid \beta_2, \beta_3, \beta_4, \beta_0)}{MS_E} = 509.36$$

$$f_2 = \frac{SS_R(\beta_2 \mid \beta_1, \beta_3, \beta_4, \beta_0)}{MS_E} = 13.15$$

$$f_3 = \frac{SS_R(\beta_3 \mid \beta_1, \beta_2, \beta_4, \beta_0)}{MS_E} = 0.59$$

$$f_4 = \frac{SS_R(\beta_4 \mid \beta_1, \beta_2, \beta_3, \beta_0)}{MS_E} = 9.39$$

Clearly, x_3 does not contribute to the fit and should be removed. In the three-variable model with regressors (x_1, x_2, x_4), all regressors are significant by the partial F-test criterion, and, consequently, the algorithm terminates. Note that backward elimination has resulted in the same model that was found by forward selection and stepwise regression. This may not always happen.

Some Comments on Final Model Selection

We have illustrated several different approaches to the selection of variables in multiple linear regression. The final model obtained from any model-building procedure should be subjected to the usual adequacy checks, such as residual analysis, lack-of-fit testing, and examination of the effects of influential points. The analyst may also consider augmenting the original set of candidate variables with cross-products, polynomial terms, or other transformations of the original variables that might improve the model. A major criticism of variable selection methods such as stepwise regression is that the analyst may conclude there is one "best" regression equation. Generally, this is not the case, because several equally good regression models can often be used. One way to avoid this problem is to use several different model-building techniques and see if different models result. For example, we have found the same model for the soft-drink delivery time data using stepwise regression, forward selection, and backward elimination. The same model was also found from all possible regressions. This is a good indication that the three-variable model is the best regression equation.

If the number of candidate regressors is not too large, the all-possible regressions method is recommended. We usually recommend using the minimum MS_E and C_p evaluation criteria in conjunction with this procedure. The all-possible regressions approach can find the "best" regression equation with respect to these criteria, while stepwise type methods offer no such assurance. Furthermore, the all-possible regressions procedure is not distorted by dependencies among the regressors, as stepwise-type methods are.

11-11.3 Stepwise Regression Computer Output

Computer programs are an indispensable part of modern regression model building, and several excellent programs are available. Minitab will perform basic stepwise regression. We will illustrate the output from another widely used software package, the Statistical Analysis System (SAS). We will use the wire bond pull strength data in Table 11-12.

Figure 11-16 presents the output from SAS PROC REG and PROC RSQUARE. PROC REG will perform stepwise regression, forward selection, backward elimination, and several other variations of the stepwise approach to variable selection. PROC RSQUARE is an all-possible regressions algorithm. Notice that the results of these runs were used to produce the stepwise regression computations illustrated in Section 11-11.2.

EXERCISES FOR SECTION 11-11

11-54. Use the National Football League Team Performance data in Exercise 11-4 to build regression models using the following techniques:

(a) All possible regressions. Find the equations that minimize MS_E and that minimize C_p.
(b) Stepwise regression.
(c) Forward selection.
(d) Backward elimination.
(e) Comment on the various models obtained. Which model seems "best," and why?

11-55. Use the gasoline mileage data in Exercise 11-5 to build regression models using the following techniques:

(a) All possible regressions. Find the minimum C_p and minimum MS_E equations.
(b) Stepwise regression.
(c) Forward selection.
(d) Backward elimination.
(e) Comment on the various models obtained.

11-56. Consider the electric power data in Exercise 11-6. Build regression models for the data using the following techniques:

(a) All possible regressions.
(b) Stepwise regression.
(c) Forward selection.
(d) Backward elimination.
(e) Comment on the models obtained. Which model would you prefer?

11-57. Consider the wire bond pull strength data in Exercise 11-8. Build regression models for the data using the following methods:

(a) Stepwise regression.
(b) Forward selection.
(c) Backward elimination.

FORWARD SELECTION PROCEDURE FOR DEPENDENT VARIABLE Y

STEP 1 VARIABLE X1 ENTERED R SQUARE = 0.93695437 C(P) = 50.46213232

	DF	SUM OF SQUARES	MEAN SQUARE	F	PROB>F
REGRESSION	1	5885.85206919	5885.85206919	615.08	0.0001
ERROR	23	220.09263481	9.56824499		
TOTAL	24	6105.94470400			

	B VALUE	STD ERROR	TYPE II SS	F	PROB>F
INTERCEPT	5.11451557				
X1	2.90270442	0.11704072	5885.85206919	615.08	0.0001

BOUNDS ON CONDITION NUMBER: 1, 1

STEP 2 VARIABLE X4 ENTERED R SQUARE = 0.98302963 C(P) = 14.64453722

	DF	SUM OF SQUARES	MEAN SQUARE	F	PROB>F
REGRESSION	2	6002.32458234	3001.16229117	637.19	0.0001
ERROR	22	103.62012166	4.71000553		
TOTAL	24	6105.94470400			

	B VALUE	STD ERROR	TYPE II SS	F	PROB>F
INTERCEPT	2.64573312				
X1	2.54772805	0.10880288	2582.53810667	548.31	0.0001
X4	3.54854459	0.71359082	116.47251315	24.73	0.0001

BOUNDS ON CONDITION NUMBER: 1.7755752, 7.023007

STEP 3 VARIABLE X2 ENTERED R SQUARE = 0.98961493 C(P) = 3.58888771

	DF	SUM OF SQUARES	MEAN SQUARE	F	PROB>F
REGRESSION	3	6042.53402205	2014.17800735	667.04	0.0001
ERROR	21	63.41068195	3.01955628		
TOTAL	24	6105.94470400			

	B VALUE	STD ERROR	TYPE II SS	F	PROB>F
INTERCEPT	1.36706835				
X1	2.53491873	0.08718736	2552.49064020	845.32	0.0001
X2	0.00852152	0.00233520	40.20943970	13.32	0.0015
X4	2.59927756	0.62779136	51.76280083	17.14	0.0005

BOUNDS ON CONDITION NUMBER: 2.119696, 15.86207

NO OTHER VARIABLES MET THE 0.1000 SIGNIFICANCE LEVEL FOR ENTRY INTO THE MODEL.

Figure 11-16 Output from SAS PROC REG and PROC RSQUARE for the wire bond pull strength data.

SUMMARY OF FORWARD SELECTION PROCEDURE FOR DEPENDENT VARIABLE Y

STEP	VARIABLE ENTERED	NUMBER IN	PARTIAL R**2	MODEL R**2	C(P)	F	PROB>F
1	X1	1	0.9640	0.9640	50.4621	615.0801	0.0001
2	X4	2	0.0191	0.9830	14.6445	24.7287	0.0001
3	X2	3	0.0066	0.9896	3.5889	13.3163	0.0015

BACKWARD ELIMINATION PROCEDURE FOR DEPENDENT VARIABLE Y

STEP 0 ALL VARIABLES ENTERED R SQUARE = 0.98991196 C(P) = 5.00000000

	DF	SUM OF SQUARES	MEAN SQUARE	F	PROB>F
REGRESSION	4	6044.34770775	1511.08692694	490.64	0.0001
ERROR	20	61.59699625	3.07984981		
TOTAL	24	6105.94470400			

	B VALUE	STD ERROR	TYPE II SS	F	PROB>F
INTERCEPT	1.06813405				
X1	2.48423244	0.11007295	1568.74664163	509.36	0.0001
X2	0.00855369	0.00235877	40.50082723	13.15	0.0017
X3	0.62236914	0.81102045	1.81368571	0.59	0.4518
X4	2.29392355	0.74854920	28.92322543	9.39	0.0061

BOUNDS ON CONDITION NUMBER: 3.980888, 44.3724

STEP 1 VARIABLE X3 REMOVED R SQUARE = 0.98961493 C(P) = 3.58888771

	DF	SUM OF SQUARES	MEAN SQUARE	F	PROB>F
REGRESSION	3	6042.53402205	2014.17800735	667.04	0.0001
ERROR	21	63.41068195	3.01955628		
TOTAL	24	6105.94470400			

	B VALUE	STD ERROR	TYPE II SS	F	PROB>F
INTERCEPT	1.36706835				
X1	2.53491873	0.08718736	2552.49064020	845.32	0.0001
X2	0.00852152	0.00233520	40.20943970	13.32	0.0015
X4	2.59927756	0.62779136	51.76280083	17.14	0.0005

BOUNDS ON CONDITION NUMBER: 2.119696, 15.86207

ALL VARIABLES IN THE MODEL ARE SIGNIFICANT AT THE 0.1000 LEVEL.

SUMMARY OF BACKWARD ELIMINATION PROCEDURE FOR DEPENDENT VARIABLE Y

STEP	VARIABLE REMOVED	NUMBER IN	PARTIAL R**2	MODEL R**2	C(P)	F	PROB>F
1	X3	3	0.0003	0.9896	3.5889	0.5889	0.4518

Figure 11-16 *(Continued)*

STEPWISE REGRESSION PROCEDURE FOR DEPENDENT VARIABLE Y

NOTE: SLSTAY HAS BEEN SET TO .15 FOR THE STEPWISE TECHNIQUE.

STEP 1 VARIABLE X1 ENTERED R SQUARE = 0.96395437 C(P) = 50.46213232

	DF	SUM OF SQUARES	MEAN SQUARE	F	PROB>F
REGRESSION	1	5885.85206919	5885.85206919	615.08	0.0001
ERROR	23	220.09263481	9.56824499		
TOTAL	24	6105.94470400			

	B VALUE	STD ERROR	TYPE II SS	F	PROB>F
INTERCEPT	5.11451557				
X1	2.90270442	0.11704072	5885.85206919	615.08	0.0001

BOUNDS ON CONDITION NUMBER: 1, 1

STEP 2 VARIABLE X4 ENTERED R SQUARE = 0.98302963 C(P) = 14.64453722

	DF	SUM OF SQUARES	MEAN SQUARE	F	PROB>F
REGRESSION	2	6002.32458234	3001.16229117	637.19	0.0001
ERROR	22	103.62012166	4.71000553		
TOTAL	24	6105.94470400			

	B VALUE	STD ERROR	TYPE II SS	F	PROB>F
INTERCEPT	2.64573312				
X1	2.54772805	0.10880288	2582.53810667	548.31	0.0001
X4	3.54854459	0.71359082	116.47251315	24.73	0.0001

BOUNDS ON CONDITION NUMBER: 1.7755752, 7.023007

STEP 3 VARIABLE X2 ENTERED R SQUARE = 0.98961493 C(P) = 3.58888771

	DF	SUM OF SQUARES	MEAN SQUARE	F	PROB>F
REGRESSION	3	6042.53402205	2014.17800735	667.04	0.0001
ERROR	21	63.41068195	3.01955628		
TOTAL	24	6105.94470400			

	B VALUE	STD ERROR	TYPE II SS	F	PROB>F
INTERCEPT	1.36706835				
X1	2.53491873	0.08718736	2552.49064020	845.32	0.0001
X2	0.00852152	0.00233520	40.20943970	13.32	0.0015
X4	2.59927756	0.62779136	51.76280083	17.14	0.0005

BOUNDS ON CONDITION NUMBER: 2.119696, 15.86207

NO OTHER VARIABLES MET THE 0.1000 SIGNIFICANCE LEVEL FOR ENTRY INTO THE MODEL.

Figure 11-16 *(Continued)*

SUMMARY OF STEPWISE REGRESSION PROCEDURE FOR DEPENDENT VARIABLE Y

STEP	VARIABLE ENTERED	VARIABLE REMOVED	NUMBER IN	PARTIAL R**2	MODEL R**2	C(P)	F	PROB>F
1	X1		1	0.9640	0.9640	50.4621	615.0801	0.0001
2	X4		2	0.0191	0.9830	14.6445	24.7287	0.0001
3	X2		3	0.0066	0.9896	3.5889	13.3163	0.0015

N = 25 REGRESSION MODELS FOR DEPENDENT VARIABLE: Y MODEL: MODEL 1

NUMBER IN MODEL	R-SQUARE	C(P)	VARIABLES IN MODEL			
1	0.24291747	1479.951	X2			
1	0.56007492	851.172	X4			
1	0.69830969	577.115	X3			
1	0.96395437	50.462132	X1			
2	0.57158123	830.360	X2	X4		
2	0.72162460	532.892	X2	X3		
2	0.72232683	531.500	X3	X4		
2	0.97220237	36.110087	X1	X3		
2	0.98113748	18.395811	X1	X2		
2	0.98302963	14.644537	X1	X4		
3	0.73299076	512.358	X2	X3	X4	
3	0.98327895	16.150260	X1	X3	X4	
3	0.98517507	12.391116	X1	X2	X3	
3	0.98961493	3.588888	X1	X2	X4	
4	0.98991196	5.000000	X1	X2	X3	X4

Figure 11-16 (*Continued*)

(d) Comment on the models obtained. Which model would you prefer?

11-58. Consider the NHL data in Exercise 11-11. Build regression models for these data using the following methods:

(a) Stepwise regression.
(b) Forward selection.
(c) Backward elimination.
(d) Which model would you prefer?

11-59. Consider the data in Exercise 11-51. Use all the terms in the full quadratic model as the candidate regressors.

(a) Use forward selection to identify a model.
(b) Use backward elimination to identify a model.
(c) Compare the two models obtained in parts (a) and (b) above. Which model would you prefer and why?

11-60. Find the minimum C_p equation and the equation that maximizes the adjusted R^2 statistic for the wire bond pull strength data in Exercise 11-8. Does the same equation satisfy both criteria?

11-61. For the NHL data in Exercise 11-11,

(a) Find the equation that minimizes C_p.

(b) Find the equation that minimizes MS_E.

(c) Find the equation that maximizes the adjusted R^2. Is this the same equation you found in part (b)?

11-62. We have used a sample of 30 observations to fit a regression model. The full model has nine regressors, the variance estimate is $\hat{\sigma}^2 = MS_E = 100$, and $R^2 = 0.92$.

(a) Calculate the F-statistic for testing significance of regression. Using $\alpha = 0.05$, what would you conclude?

(b) Suppose that we fit another model using only four of the original regressors and that the error sum of squares for this new model is 2200. Find the estimate of σ^2 for this new reduced model. Would you conclude that the reduced model is superior to the old one? Why?

(c) Find the value of C_p for the reduced model in part (b). Would you conclude that the reduced model is better than the old model?

11-63. A sample of 25 observations is used to fit a regression model in seven variables. The estimate of σ^2 for this full model is $MS_E = 10$.

(a) A forward selection algorithm has put three of the original seven regressors in the model. The error sum of squares for the three-variable model is $SS_E = 300$. Based on C_p, would you conclude that the three-variable model has any remaining bias?

(b) After looking at the forward selection model in part (a), suppose you could add one more regressor to the model. This regressor will reduce the error sum of squares to 275. Will the addition of this variable improve the model? Why?

11-12 MULTICOLLINEARITY

In multiple regression problems, we expect to find dependencies between the response variable Y and the regressors x_j. In most regression problems, however, we find that there are also dependencies among the regressor variables x_j. In situations where these dependencies are strong, we say that **multicollinearity** exists. Multicollinearity can have serious effects on the estimates of the regression coefficients and on the general applicability of the estimated model.

The effects of multicollinearity may be easily demonstrated. The diagonal elements of the matrix $\mathbf{C} = (\mathbf{X}'\mathbf{X})^{-1}$ can be written as

$$C_{jj} = \frac{1}{(1 - R_j^2)} \qquad j = 1, 2, \ldots, k$$

where R_j^2 is the coefficient of multiple determination resulting from regressing x_j on the other $k - 1$ regressor variables. Clearly, the stronger the linear dependency of x_j on the remaining regressor variables, and hence the stronger the multicollinearity, the larger the value of R_j^2 will be. Recall that $V(\hat{\beta}_j) = \sigma^2 C_{jj}$. Therefore, we say that the variance of $\hat{\beta}_j$ is "inflated" by the quantity $(1 - R_j^2)^{-1}$. Consequently, we define the **variance inflation factor** for $\hat{\beta}_j$ as

$$VIF(\hat{\beta}_j) = \frac{1}{(1 - R_j^2)} \qquad j = 1, 2, \ldots, k \tag{11-50}$$

These factors are an important measure of the extent to which multicollinearity is present.

Although the estimates of the regression coefficients are very imprecise when multicollinearity is present, the fitted model equation may still be useful. For example, suppose we wish to predict new observations on the response. If these predictions are interpolations in the original region of the x-space where the multicollinearity is in effect, then satisfactory predictions will often be obtained, because while individual β_j may be poorly estimated, the function $\sum_{j=1}^{k} \beta_j x_{ij}$ may be estimated quite well. On the other hand, if the prediction of new observations requires extrapolation beyond the original region of the x-space where the data were collected, then generally we would expect to obtain poor results. Extrapolation usually requires good estimates of the individual model parameters.

Multicollinearity arises for several reasons. It will occur when the analyst collects data such that a linear constraint holds approximately among the columns of the $\mathbf{X}$ matrix. For example, if four regressor variables are the components of a mixture, then such a constraint will always exist because the sum of the components is always constant. Usually, these constraints do not hold exactly, and the analyst might not know that they exist.

The presence of multicollinearity can be detected in several ways. Two of the more easily understood of these will be discussed briefly.

1. The **variance inflation factors**, defined in equation 11-50, are very useful measures of multicollinearity. The larger the variance inflation factor, the more severe the multicollinearity. Some authors have suggested that if any variance inflation factor exceeds 10, then multicollinearity is a problem. Other authors consider this value too liberal and suggest that the variance inflation factors should not exceed 4 or 5. Minitab will calculate the variance inflation factors. Table 11-4 presents the Minitab multiple regression output for the wire bond pull strength data. Since both VIF_1 and VIF_2 are small, there is no problem with multicollinearity.
2. If the F-test for significance of regression is significant, but tests on the individual regression coefficients are not significant, then multicollinearity may be present.

Several remedial measures have been proposed for solving the problem of multicollinearity. Augmenting the data with new observations specifically designed to break up the approximate linear dependencies that currently exist is often suggested. However, this is sometimes impossible because of economic reasons or because of the physical constraints that relate the x_j. Another possibility is to delete certain variables from the model, but this approach has the disadvantage of discarding the information contained in the deleted variables.

Since multicollinearity primarily affects the stability of the regression coefficients, it would seem that estimating these parameters by some method that is less sensitive to multicollinearity than ordinary least squares would be helpful. Several methods have been suggested. One alternative to ordinary least squares, **ridge regression,** can be useful in combating multicollinearity. For more details on ridge regression, see Montgomery and Peck (1992) and Myers (1990).

Supplemental Exercises

11-64. The data shown in the table on page 557 represent the thrust of a jet-turbine engine (y) and six candidate regressors: x_1 = primary speed of rotation, x_2 = secondary speed of rotation, x_3 = fuel flow rate, x_4 = pressure, x_5 = exhaust temperature, and x_6 = ambient temperature at time of test.

(a) Fit a multiple linear regression model using x_3 = fuel flow rate, x_4 = pressure, and x_5 = exhaust temperature as the regressors.

(b) Test for significance of regression using $\alpha = 0.01$. Find the P-value for this test. What are your conclusions?

(c) Find the t-test statistic for each regressor. Using $\alpha = 0.01$, explain carefully the conclusion you can draw from these statistics.

(d) Find R^2 and the adjusted statistic for this model.

(e) Construct a normal probability plot of the residuals and interpret this graph.

(f) Plot the residuals versus $\hat{y}$. Are there any indications of inequality of variance or nonlinearity?

(g) Plot the residuals versus x_3. Is there any indication of nonlinearity?

(h) Predict the thrust for an engine for which $x_3 = 1670$, $x_4 = 170$, and $x_5 = 1589$.

11-65. Consider the engine thrust data in Exercise 11-64. Refit the model using $y^* = \ln y$ as the response variable and $x_3^* = \ln x_3$ as the regressor (along with x_4 and x_5).

(a) Test for significance of regression using $\alpha = 0.01$. Find the P-value for this test and state your conclusions.

(b) Use the t-statistic to test H_0: $\beta_j = 0$ versus H_1: $\beta_j \neq 0$ for each variable in the model. If $\alpha = 0.01$, what conclusions can you draw?

(c) Plot the residuals versus $\hat{y}^*$ and versus x_3^*. Comment on these plots. How do they compare with their counterparts obtained in Exercise 11-64 parts (f) and (g)?

11-66. The transient points of an electronic inverter are influenced by many factors. On page 558 are data on the transient point (y, in volts) of PMOS-NMOS inverters and five candidate regressors:, x_1 = width of the NMOS device, x_2 = length of the NMOS device, x_3 = width of the PMOS device, x_4 = length of the PMOS device, and x_5 = temperature (°C).

(a) Fit the multiple linear regression model to these data. Test for significance of regression using $\alpha = 0.01$. Find the P-value for this test and use it to draw your conclusions.

(b) Test the contribution of each variable to the model using the t-test with $\alpha = 0.05$. What are your conclusions?

(c) Delete x_5 from the model. Test the new model for significance of regression. Also test the relative contribution of each regressor to the new model with the t-test. Using $\alpha = 0.05$, what are your conclusions?

Observation Number	y	x_1	x_2	x_3	x_4	x_5	x_6
1	4540	2140	20640	30250	205	1732	99
2	4315	2016	20280	30010	195	1697	100
3	4095	1905	19860	29780	184	1662	97
4	3650	1675	18980	29330	164	1598	97
5	3200	1474	18100	28960	144	1541	97
6	4833	2239	20740	30083	216	1709	87
7	4617	2120	20305	29831	206	1669	87
8	4340	1990	19961	29604	196	1640	87
9	3820	1702	18916	29088	171	1572	85
10	3368	1487	18012	28675	149	1522	85
11	4445	2107	20520	30120	195	1740	101
12	4188	1973	20130	29920	190	1711	100
13	3981	1864	19780	29720	180	1682	100
14	3622	1674	19020	29370	161	1630	100
15	3125	1440	18030	28940	139	1572	101
16	4560	2165	20680	30160	208	1704	98
17	4340	2048	20340	29960	199	1679	96
18	4115	1916	19860	29710	187	1642	94
19	3630	1658	18950	29250	164	1576	94
20	3210	1489	18700	28890	145	1528	94
21	4330	2062	20500	30190	193	1748	101
22	4119	1929	20050	29960	183	1713	100
23	3891	1815	19680	29770	173	1684	100
24	3467	1595	18890	29360	153	1624	99
25	3045	1400	17870	28960	134	1569	100
26	4411	2047	20540	30160	193	1746	99
27	4203	1935	20160	29940	184	1714	99
28	3968	1807	19750	29760	173	1679	99
29	3531	1591	18890	29350	153	1621	99
30	3074	1388	17870	28910	133	1561	99
31	4350	2071	20460	30180	198	1729	102
32	4128	1944	20010	29940	186	1692	101
33	3940	1831	19640	29750	178	1667	101
34	3480	1612	18710	29360	156	1609	101
35	3064	1410	17780	28900	136	1552	101
36	4402	2066	20520	30170	197	1758	100
37	4180	1954	20150	29950	188	1729	99
38	3973	1835	19750	29740	178	1690	99
39	3530	1616	18850	29320	156	1616	99
40	3080	1407	17910	28910	137	1569	100

Observation Number	x_1	x_2	x_3	x_4	x_5	y
1	3	3	3	3	0	0.787
2	8	30	8	8	0	0.293
3	3	6	6	6	0	1.710
4	4	4	4	12	0	0.203
5	8	7	6	5	0	0.806
6	10	20	5	5	0	4.713
7	8	6	3	3	25	0.607
8	6	24	4	4	25	9.107
9	4	10	12	4	25	9.210
10	16	12	8	4	25	1.365
11	3	10	8	8	25	4.554
12	8	3	3	3	25	0.293
13	3	6	3	3	50	2.252
14	3	8	8	3	50	9.167
15	4	8	4	8	50	0.694
16	5	2	2	2	50	0.379
17	2	2	2	3	50	0.485
18	10	15	3	3	50	3.345
19	15	6	2	3	50	0.208
20	15	6	2	3	75	0.201
21	10	4	3	3	75	0.329
22	3	8	2	2	75	4.966
23	6	6	6	4	75	1.362
24	2	3	8	6	75	1.515
25	3	3	8	8	75	0.751

(d) Notice that the MS_E for the model in part (c) is smaller than the MS_E for the full model in part (a). Explain why this has occurred.

(e) Calculate the studentized residuals. Do any of these seem unusually large?

(f) Suppose that you learn that the second observation was incorrectly recorded. Delete this observation and refit the model using x_1, x_2, x_3, and x_4 as the regressors. Notice that the R^2 for this model is considerably higher than the R^2 for either of the models fitted previously. Explain why the R^2 for this model has increased.

(g) Test the model from part (f) for significance of regression using $\alpha = 0.05$. Also investigate the contribution of each regressor to the model using the t-test with $\alpha = 0.05$. What conclusions can you draw?

(h) Plot the residuals from the model in part (f) versus $\hat{y}$ and versus each of the regressors x_1, x_2, x_3, and x_4. Comment on the plots.

11-67. Consider the inverter data in Exercise 11-66. Delete observation 2 from the original data. Define new variables as follows: $y^* = ln\ y$, $x_1^* = 1/\sqrt{x_1}$, $x_2^* = \sqrt{x_2}$, $x_3^* = 1/\sqrt{x_3}$, and $x_4^* = \sqrt{x_4}$.

(a) Fit a regression model using these transformed regressors (do not use x_5).

(b) Test the model for significance of regression using $\alpha = 0.05$. Use the t-test to investigate the contribution of each variable to the model ($\alpha = 0.05$). What are your conclusions?

(c) Plot the residuals versus $\hat{y}^*$ and versus each of the transformed regressors. Comment on the plots.

11-68. Below are data on y = green liquor (g/l) and x = paper machine speed (ft/min) from a Kraft paper machine. (The data were read from a graph in an article in the *Tappi Journal,* March 1986.)

y	16.0	15.8	15.6	15.5	14.8
x	1700	1720	1730	1740	1750
y	14.0	13.5	13.0	12.0	11.0
x	1760	1770	1780	1790	1795

(a) Fit the model $Y = \beta_0 + \beta_1 x + \beta_2 x^2 + \epsilon$ using least squares.

(b) Test for significance of regression using $\alpha = 0.05$. What are your conclusions?

(c) Test the contribution of the quadratic term to the model, over the contribution of the linear term, using an F-statistic. If $\alpha = 0.05$, what conclusion can you draw?

(d) Plot the residuals from the model

in part (a) versus $\hat{y}$. Does the plot reveal any inadequacies?

(e) Construct a normal probability plot of the residuals. Comment on the normality assumption.

11-69. Consider the jet engine thrust data in Exercise 11-64.

(a) Use all possible regressions to select the best regression equation, where the model with the minimum value of MS_E is to be selected as "best."

(b) Repeat part (a) using the C_p criterion to identify the best equation.

(c) Use stepwise regression to select a subset regression model.

(d) Compare the models obtained in parts (a), (b), and (c) above.

11-70. Consider the electronic inverter data in Exercise 11-66 and 11-67. Define the response and regressors variables as in Exercise 11-67, and delete the second observation in the sample.

(a) Use all possible regressions to find the equation that minimizes C_p.

(b) Use all possible regressions to find the equation that minimizes MS_E.

(c) Use stepwise regression to select a subset regression model.

(d) Compare the models you have obtained.

11-71. Consider the three-variable regression model for the jet engine thrust data in Exercise 11-65. Calculate the variance inflation factors for this model. Would you conclude that multicollinearity is a problem in this model?

11-72. A multiple regression model was used to relate y = viscosity of a chemical product to x_1 = temperature and x_2 = reaction time. The data set consisted of $n = 15$ observations.

(a) The estimated regression coefficients were $\hat{\beta}_0 = 300.00$, $\hat{\beta}_1 = 0.85$, and $\hat{\beta}_2 = 10.40$. Calculate an estimate of mean viscosity when $x_1 = 100°F$ and $x_2 = 2$ hr.

(b) The sums of squares were $SS_T = 1230.50$ and $SS_E = 120.30$. Test for significance of regression using $\alpha = 0.05$. What conclusion can you draw?

(c) What proportion of total variability in viscosity is accounted for by the variables in this model?

(d) Suppose that another regressor, x_3 = stirring rate, is added to the model. The new value of the error sum of squares is $SS_E = 117.20$. Has adding the new variable resulted in a smaller value of MS_E? Discuss the significance of this result.

(e) Calculate an F-statistic to assess the contribution of x_3 to the model. Using $\alpha = 0.05$, what conclusions do you reach?

11-73. An article in the *Journal of the American Ceramics Society* (Vol. 75, 1992, pp. 112–116) describes a process for immobilizing chemical or nuclear wastes in soil by dissolving the contaminated soil into a glass block. The authors mix CaO and Na_2O with soil and model viscosity and electrical conductivity. The electrical conductivity model involves six regressors, and the sample consists of $n = 14$ observations.

(a) For the six-regressor model, suppose that $SS_T = 0.50$ and $R^2 = 0.94$. Find SS_E and SS_R, and use this information to test for significance of regression with $\alpha = 0.05$. What are your conclusions?

(b) Suppose that one of the original regressors is deleted from the model, resulting in $R^2 = 0.92$. What can you conclude about the contribution of the variable that was removed? Answer this question by calculating an F-statistic.

(c) Does deletion of the regressor variable in part (b) result in a smaller value of MS_E for the five-variable model, in comparison to the original six-variable model? Comment on the significance of your answer.

MIND-EXPANDING EXERCISES

11-74. Consider a multiple regression model with k regressors. Show that the test statistic for significance of regression can be written as

$$F_0 = \frac{R^2/k}{(1 - R^2)/(n - k - 1)}$$

Suppose that $n = 20$, $k = 4$, and $R^2 = 0.90$. If $\alpha = 0.05$, what conclusion would you draw about the relationship between y and the four regressors?

11-75. A regression model is used to relate a response y to $k = 4$ regressors with $n = 20$. What is the smallest value of R^2 that will result in a significant regression if $\alpha = 0.05$? Use the results of the previous exercise. Are you surprised by how small the value of R^2 is?

11-76. Show that we can express the residuals from a multiple regression model as $\mathbf{e} = (\mathbf{I} - \mathbf{H})\mathbf{y}$, where $\mathbf{H} = \mathbf{X}(\mathbf{X}'\mathbf{X})^{-1}\mathbf{X}'$.

11-77. Show that the variance of the ith residual e_i in a multiple regression model is $\sigma(1 - h_{ii})$ and that the covariance between e_i and e_j is $-\sigma^2 h_{ij}$, where the h's are the elements of $\mathbf{H} = \mathbf{X}(\mathbf{X}'\mathbf{X})^{-1}\mathbf{X}'$.

11-78. Consider the multiple linear regression model $\mathbf{y} = \mathbf{X}\boldsymbol{\beta} + \boldsymbol{\epsilon}$. If $\hat{\boldsymbol{\beta}}$ denotes the least squares estimator of $\boldsymbol{\beta}$, show that $\hat{\boldsymbol{\beta}} = \boldsymbol{\beta} + \mathbf{R}\boldsymbol{\epsilon}$, where $\mathbf{R} = (\mathbf{X}'\mathbf{X})^{-1}\mathbf{X}'$.

11-79. **Constrained Least Squares.** Suppose we wish to find the least squares estimator of $\boldsymbol{\beta}$ in the model $\mathbf{y} = \mathbf{X}\boldsymbol{\beta} + \boldsymbol{\epsilon}$ subject to a set of equality constraints, say, $\mathbf{T}\boldsymbol{\beta} = \mathbf{c}$.

(a) Show that the estimator is

$$\hat{\boldsymbol{\beta}}_c = \hat{\boldsymbol{\beta}} + (\mathbf{X}'\mathbf{X})^{-1} \times \mathbf{T}'[\mathbf{T}(\mathbf{X}'\mathbf{X})^{-1}\mathbf{T}']^{-1}(\mathbf{c} - \mathbf{T}\hat{\boldsymbol{\beta}})$$

where $\hat{\boldsymbol{\beta}} = (\mathbf{X}'\mathbf{X})^{-1}\mathbf{X}'\mathbf{y}$.

(b) Discuss situations where this model might be appropriate.

11-80. **Piecewise Linear Regression (I).** Suppose that y is piecewise linearly related to x. That is, different linear relationships are appropriate over the intervals $-\infty < x \le x^*$ and $x^* < x < \infty$. Show how indicator variables can be used to fit such a piecewise linear regression model, assuming that the point x^* is known.

11-81. **Piecewise Linear Regression (II).** Consider the piecewise linear regression model described in Exercise 11-80. Suppose that at the point x^* a discontinuity occurs in the regression function. Show how indicator variables can be used to incorporate the discontinuity into the model.

11-82. **Piecewise Linear Regression (III).** Consider the piecewise linear regression model described in Exercise 11-80. Suppose that the point x^* is not known with certainty and must be estimated. Suggest an approach that could be used to fit the piecewise linear regression model.

APPENDICES

APPENDIX A. STATISTICAL TABLES AND CHARTS A-1

APPENDIX B. SUPPLEMENTAL TECHNICAL MATERIAL B-1

APPENDIX C. BIBLIOGRAPHY C-1

APPENDIX D. ANSWERS TO SELECTED EXERCISES D-1

Appendix A Statistical Tables and Charts

Table I Summary of Common Probability Distributions

Name	Probability Density Function	Mean	Variance	Section in Book
Discrete				
Uniform	$\frac{1}{b-a}, a \le b$	$\frac{(b+a)}{2}$	$\frac{(b-a+1)^2-1}{12}$	4-5
Binomial	$\binom{n}{x} p^x(1-p)^{n-x}$, $x = 0, 1, \ldots, n, 0 \le p \le 1$	np	$np(1-p)$	4-6
Geometric	$(1-p)^{x-1}p$, $x = 1, 2, \ldots, 0 \le p \le 1$	$1/p$	$(1-p)/p^2$	4-7.1
Negative binomial	$\binom{x-1}{r-1}(1 \le p)^{x-r}p^r$ $x = r, r+1, r+2, \ldots,$ $0 \le p \le 1$	r/p	$r(1-p)/p^2$	4-7.2
Hypergeometric	$\frac{\binom{K}{x}\binom{N-k}{n-x}}{\binom{N}{n}}$ $x = \max(0, n-N+K), 1, \ldots$ $\min(K, n), K \le N, n \le N$	np, where $p = \frac{K}{N}$	$np(1-p)\left(\frac{N-m}{N-1}\right)$	4-8
Poisson	$\frac{e^{-\lambda}\lambda^x}{x!}, x = 0, 1, 2, \ldots, \lambda > 0$	λ	λ	4-9
Continuous				
Uniform	$\frac{1}{b-a}, a \le x \le b$	$\frac{(b+a)}{2}$	$\frac{(b-a)^2}{12}$	5-5
Normal	$\frac{1}{\sqrt{2\pi}\sigma} e^{\frac{-(x-\mu)^2}{2\sigma^2}}$, $-\infty < x < \infty, -\infty < \mu < \infty, \sigma > 0$	μ	σ^2	5-6
Exponential	$\lambda e^{-\lambda x}, x \ge 0 \quad \lambda > 0$	$1/\lambda$	$1/\lambda^2$	5-9
Erlang	$\frac{\lambda^r x^{r-1} e^{-\lambda x}}{(r-1)!}, x > 0, r = 1, 2, \ldots$	r/λ	r/λ^2	5-10.1
Gamma	$\frac{\lambda x^{r-1} e^{-\lambda x}}{\Gamma(r)}, x > 0, r > 0, \lambda > 0$	r/λ	r/λ^2	5-10.2
Weibull	$\frac{\beta}{\delta}\left(\frac{x}{\delta}\right)^{\beta-1} e^{-(x/\delta)^\beta}$, $x > 0, \beta > 0, \delta > 0$	$\delta\Gamma\left(1+\frac{1}{\beta}\right)$	$\delta^2\Gamma\left(1+\frac{2}{\beta}\right) - \delta^2\left[\Gamma\left(1+\frac{1}{\beta}\right)\right]^2$	5-11

$$\Phi(z) = P(Z \leq z) = \int_{-\infty}^{z} \frac{1}{\sqrt{2\pi}} e^{\frac{-u^2}{2}} du$$

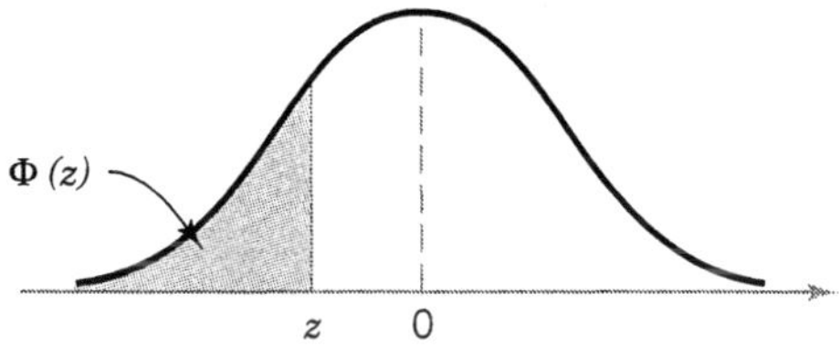

Table II Cumulative Standard Normal Distribution

z	−0.09	−0.08	−0.07	−0.06	−0.05	−0.04	−0.03	−0.02	−0.01	−0.00
−3.9	0.000033	0.000034	0.000036	0.000037	0.000039	0.000041	0.000042	0.000044	0.000046	0.000048
−3.8	0.000050	0.000052	0.000054	0.000057	0.000059	0.000062	0.000064	0.000067	0.000069	0.000072
−3.7	0.000075	0.000078	0.000082	0.000085	0.000088	0.000092	0.000096	0.000100	0.000104	0.000108
−3.6	0.000112	0.000117	0.000121	0.000126	0.000131	0.000136	0.000142	0.000147	0.000153	0.000159
−3.5	0.000165	0.000172	0.000179	0.000185	0.000193	0.000200	0.000208	0.000216	0.000224	0.000233
−3.4	0.000242	0.000251	0.000260	0.000270	0.000280	0.000291	0.000302	0.000313	0.000325	0.000337
−3.3	0.000350	0.000362	0.000376	0.000390	0.000404	0.000419	0.000434	0.000450	0.000467	0.000483
−3.2	0.000501	0.000519	0.000538	0.000557	0.000577	0.000598	0.000619	0.000641	0.000664	0.000687
−3.1	0.000711	0.000736	0.000762	0.000789	0.000816	0.000845	0.000874	0.000904	0.000935	0.000968
−3.0	0.001001	0.001035	0.001070	0.001107	0.001144	0.001183	0.001223	0.001264	0.001306	0.001350
−2.9	0.001395	0.001441	0.001489	0.001538	0.001589	0.001641	0.001695	0.001750	0.001807	0.001866
−2.8	0.001926	0.001988	0.002052	0.002118	0.002186	0.002256	0.002327	0.002401	0.002477	0.002555
−2.7	0.002635	0.002718	0.002803	0.002890	0.002980	0.003072	0.003167	0.003264	0.003364	0.003467
−2.6	0.003573	0.003681	0.003793	0.003907	0.004025	0.004145	0.004269	0.004396	0.004527	0.004661
−2.5	0.004799	0.004940	0.005085	0.005234	0.005386	0.005543	0.005703	0.005868	0.006037	0.006210
−2.4	0.006387	0.006569	0.006756	0.006947	0.007143	0.007344	0.007549	0.007760	0.007976	0.008198
−2.3	0.008424	0.008656	0.008894	0.009137	0.009387	0.009642	0.009903	0.010170	0.010444	0.010724
−2.2	0.011011	0.011304	0.011604	0.011911	0.012224	0.012545	0.012874	0.013209	0.013553	0.013903
−2.1	0.014262	0.014629	0.015003	0.015386	0.015778	0.016177	0.016586	0.017003	0.017429	0.017864
−2.0	0.018309	0.018763	0.019226	0.019699	0.020182	0.020675	0.021178	0.021692	0.022216	0.022750
−1.9	0.023295	0.023852	0.024419	0.024998	0.025588	0.026190	0.026803	0.027429	0.028067	0.028717
−1.8	0.029379	0.030054	0.030742	0.031443	0.032157	0.032884	0.033625	0.034379	0.035148	0.035930
−1.7	0.036727	0.037538	0.038364	0.039204	0.040059	0.040929	0.041815	0.042716	0.043633	0.044565
−1.6	0.045514	0.046479	0.047460	0.048457	0.049471	0.050503	0.051551	0.052616	0.053699	0.054799
−1.5	0.055917	0.057053	0.058208	0.059380	0.060571	0.061780	0.063008	0.064256	0.065522	0.066807
−1.4	0.068112	0.069437	0.070781	0.072145	0.073529	0.074934	0.076359	0.077804	0.079270	0.080757
−1.3	0.082264	0.083793	0.085343	0.086915	0.088508	0.090123	0.091759	0.093418	0.095098	0.096801
−1.2	0.098525	0.100273	0.102042	0.103835	0.105650	0.107488	0.109349	0.111233	0.113140	0.115070
−1.1	0.117023	0.119000	0.121001	0.123024	0.125072	0.127143	0.129238	0.131357	0.133500	0.135666
−1.0	0.137857	0.140071	0.142310	0.144572	0.146859	0.149170	0.151505	0.153864	0.156248	0.158655
−0.9	0.161087	0.163543	0.166023	0.168528	0.171056	0.173609	0.176185	0.178786	0.181411	0.184060
−0.8	0.186733	0.189430	0.192150	0.194894	0.197662	0.200454	0.203269	0.206108	0.208970	0.211855
−0.7	0.214764	0.217695	0.220650	0.223627	0.226627	0.229650	0.232695	0.235762	0.238852	0.241964
−0.6	0.245097	0.248252	0.251429	0.254627	0.257846	0.261086	0.264347	0.267629	0.270931	0.274253
−0.5	0.277595	0.280957	0.284339	0.287740	0.291160	0.294599	0.298056	0.301532	0.305026	0.308538
−0.4	0.312067	0.315614	0.319178	0.322758	0.326355	0.329969	0.333598	0.337243	0.340903	0.344578
−0.3	0.348268	0.351973	0.355691	0.359424	0.363169	0.366928	0.370700	0.374484	0.378281	0.382089
−0.2	0.385908	0.389739	0.393580	0.397432	0.401294	0.405165	0.409046	0.412936	0.416834	0.420740
−0.1	0.424655	0.428576	0.432505	0.436441	0.440382	0.444330	0.448283	0.452242	0.456205	0.460172
0.0	0.464144	0.468119	0.472097	0.476078	0.480061	0.484047	0.488033	0.492022	0.496011	0.500000

$$\Phi(z) = P(Z \leq z) = \int_{-\infty}^{z} \frac{1}{\sqrt{2\pi}} e^{\frac{-u^2}{2}} du$$

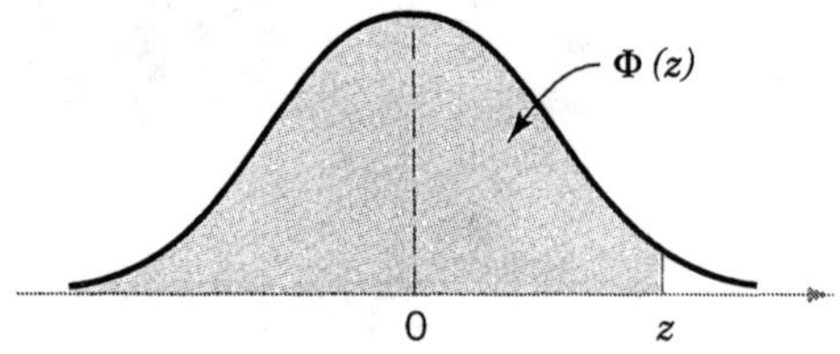

Table II Cumulative Standard Normal Distribution (*continued*)

z	0.00	0.01	0.02	0.03	0.04	0.05	0.06	0.07	0.08	0.09
0.0	0.500000	0.503989	0.507978	0.511967	0.515953	0.519939	0.532922	0.527903	0.531881	0.535856
0.1	0.539828	0.543795	0.547758	0.551717	0.555760	0.559618	0.563559	0.567495	0.571424	0.575345
0.2	0.579260	0.583166	0.587064	0.590954	0.594835	0.598706	0.602568	0.606420	0.610261	0.614092
0.3	0.617911	0.621719	0.625516	0.629300	0.633072	0.636831	0.640576	0.644309	0.648027	0.651732
0.4	0.655422	0.659097	0.662757	0.666402	0.670031	0.673645	0.677242	0.680822	0.684386	0.687933
0.5	0.691462	0.694974	0.698468	0.701944	0.705401	0.708840	0.712260	0.715661	0.719043	0.722405
0.6	0.725747	0.729069	0.732371	0.735653	0.738914	0.742154	0.745373	0.748571	0.751748	0.754903
0.7	0.758036	0.761148	0.764238	0.767305	0.770350	0.773373	0.776373	0.779350	0.782305	0.785236
0.8	0.788145	0.791030	0.793892	0.796731	0.799546	0.802338	0.805106	0.807850	0.810570	0.813267
0.9	0.815940	0.818589	0.821214	0.823815	0.826391	0.828944	0.831472	0.833977	0.836457	0.838913
1.0	0.841345	0.843752	0.846136	0.848495	0.850830	0.853141	0.855428	0.857690	0.859929	0.862143
1.1	0.864334	0.866500	0.868643	0.870762	0.872857	0.874928	0.876976	0.878999	0.881000	0.882977
1.2	0.884930	0.886860	0.888767	0.890651	0.892512	0.894350	0.896165	0.897958	0.899727	0.901475
1.3	0.903199	0.904902	0.906582	0.908241	0.909877	0.911492	0.913085	0.914657	0.916207	0.917736
1.4	0.919243	0.920730	0.922196	0.923641	0.925066	0.926471	0.927855	0.929219	0.930563	0.931888
1.5	0.933193	0.934478	0.935744	0.936992	0.938220	0.939429	0.940620	0.941792	0.942947	0.944083
1.6	0.945201	0.946301	0.947384	0.948449	0.949497	0.950529	0.951543	0.952540	0.953521	0.954486
1.7	0.955435	0.956367	0.957284	0.958185	0.959071	0.959941	0.960796	0.961636	0.962462	0.963273
1.8	0.964070	0.964852	0.965621	0.966375	0.967116	0.967843	0.968557	0.969258	0.969946	0.970621
1.9	0.971283	0.971933	0.972571	0.973197	0.973810	0.974412	0.975002	0.975581	0.976148	0.976705
2.0	0.977250	0.977784	0.978308	0.978822	0.979325	0.979818	0.980301	0.980774	0.981237	0.981691
2.1	0.982136	0.982571	0.982997	0.983414	0.983823	0.984222	0.984614	0.984997	0.985371	0.985738
2.2	0.986097	0.986447	0.986791	0.987126	0.987455	0.987776	0.988089	0.988396	0.988696	0.988989
2.3	0.989276	0.989556	0.989830	0.990097	0.990358	0.990613	0.990863	0.991106	0.991344	0.991576
2.4	0.991802	0.992024	0.992240	0.992451	0.992656	0.992857	0.993053	0.993244	0.993431	0.993613
2.5	0.993790	0.993963	0.994132	0.994297	0.994457	0.994614	0.994766	0.994915	0.995060	0.995201
2.6	0.995339	0.995473	0.995604	0.995731	0.995855	0.995975	0.996093	0.996207	0.996319	0.996427
2.7	0.996533	0.996636	0.996736	0.996833	0.996928	0.997020	0.997110	0.997197	0.997282	0.997365
2.8	0.997445	0.997523	0.997599	0.997673	0.997744	0.997814	0.997882	0.997948	0.998012	0.998074
2.9	0.998134	0.998193	0.998250	0.998305	0.998359	0.998411	0.998462	0.998511	0.998559	0.998605
3.0	0.998650	0.998694	0.998736	0.998777	0.998817	0.998856	0.998893	0.998930	0.998965	0.998999
3.1	0.999032	0.999065	0.999096	0.999126	0.999155	0.999184	0.999211	0.999238	0.999264	0.999289
3.2	0.999313	0.999336	0.999359	0.999381	0.999402	0.999423	0.999443	0.999462	0.999481	0.999499
3.3	0.999517	0.999533	0.999550	0.999566	0.999581	0.999596	0.999610	0.999624	0.999638	0.999650
3.4	0.999663	0.999675	0.999687	0.999698	0.999709	0.999720	0.999730	0.999740	0.999749	0.999758
3.5	0.999767	0.999776	0.999784	0.999792	0.999800	0.999807	0.999815	0.999821	0.999828	0.999835
3.6	0.999841	0.999847	0.999853	0.999858	0.999864	0.999869	0.999874	0.999879	0.999883	0.999888
3.7	0.999892	0.999896	0.999900	0.999904	0.999908	0.999912	0.999915	0.999918	0.999922	0.999925
3.8	0.999928	0.999931	0.999933	0.999936	0.999938	0.999941	0.999943	0.999946	0.999948	0.999950
3.9	0.999952	0.999954	0.999956	0.999958	0.999959	0.999961	0.999963	0.999964	0.999966	0.999967

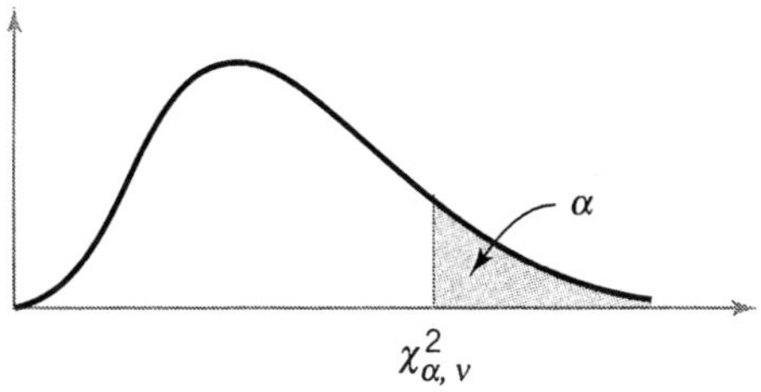

Table III Percentage Points $\chi^2_{\alpha,\nu}$ of the Chi-Squared Distribution

ν \ α	.995	.990	.975	.950	.900	.500	.100	.050	.025	.010	.005
1	.00+	.00+	.00+	.00+	.02	.45	2.71	3.84	5.02	6.63	7.88
2	.01	.02	.05	.10	.21	1.39	4.61	5.99	7.38	9.21	10.60
3	.07	.11	.22	.35	.58	2.37	6.25	7.81	9.35	11.34	12.84
4	.21	.30	.48	.71	1.06	3.36	7.78	9.49	11.14	13.28	14.86
5	.41	.55	.83	1.15	1.61	4.35	9.24	11.07	12.83	15.09	16.75
6	.68	.87	1.24	1.64	2.20	5.35	10.65	12.59	14.45	16.81	18.55
7	.99	1.24	1.69	2.17	2.83	6.35	12.02	14.07	16.01	18.48	20.28
8	1.34	1.65	2.18	2.73	3.49	7.34	13.36	15.51	17.53	20.09	21.96
9	1.73	2.09	2.70	3.33	4.17	8.34	14.68	16.92	19.02	21.67	23.59
10	2.16	2.56	3.25	3.94	4.87	9.34	15.99	18.31	20.48	23.21	25.19
11	2.60	3.05	3.82	4.57	5.58	10.34	17.28	19.68	21.92	24.72	26.76
12	3.07	3.57	4.40	5.23	6.30	11.34	18.55	21.03	23.34	26.22	28.30
13	3.57	4.11	5.01	5.89	7.04	12.34	19.81	22.36	24.74	27.69	29.82
14	4.07	4.66	5.63	6.57	7.79	13.34	21.06	23.68	26.12	29.14	31.32
15	4.60	5.23	6.27	7.26	8.55	14.34	22.31	25.00	27.49	30.58	32.80
16	5.14	5.81	6.91	7.96	9.31	15.34	23.54	26.30	28.85	32.00	34.27
17	5.70	6.41	7.56	8.67	10.09	16.34	24.77	27.59	30.19	33.41	35.72
18	6.26	7.01	8.23	9.39	10.87	17.34	25.99	28.87	31.53	34.81	37.16
19	6.84	7.63	8.91	10.12	11.65	18.34	27.20	30.14	32.85	36.19	38.58
20	7.43	8.26	9.59	10.85	12.44	19.34	28.41	31.41	34.17	37.57	40.00
21	8.03	8.90	10.28	11.59	13.24	20.34	29.62	32.67	35.48	38.93	41.40
22	8.64	9.54	10.98	12.34	14.04	21.34	30.81	33.92	36.78	40.29	42.80
23	9.26	10.20	11.69	13.09	14.85	22.34	32.01	35.17	38.08	41.64	44.18
24	9.89	10.86	12.40	13.85	15.66	23.34	33.20	36.42	39.36	42.98	45.56
25	10.52	11.52	13.12	14.61	16.47	24.34	34.28	37.65	40.65	44.31	46.93
26	11.16	12.20	13.84	15.38	17.29	25.34	35.56	38.89	41.92	45.64	48.29
27	11.81	12.88	14.57	16.15	18.11	26.34	36.74	40.11	43.19	46.96	49.65
28	12.46	13.57	15.31	16.93	18.94	27.34	37.92	41.34	44.46	48.28	50.99
29	13.12	14.26	16.05	17.71	19.77	28.34	39.09	42.56	45.72	49.59	52.34
30	13.79	14.95	16.79	18.49	20.60	29.34	40.26	43.77	46.98	50.89	53.67
40	20.71	22.16	24.43	26.51	29.05	39.34	51.81	55.76	59.34	63.69	66.77
50	27.99	29.71	32.36	34.76	37.69	49.33	63.17	67.50	71.42	76.15	79.49
60	35.53	37.48	40.48	43.19	46.46	59.33	74.40	79.08	83.30	88.38	91.95
70	43.28	45.44	48.76	51.74	55.33	69.33	85.53	90.53	95.02	100.42	104.22
80	51.17	53.54	57.15	60.39	64.28	79.33	96.58	101.88	106.63	112.33	116.32
90	59.20	61.75	65.65	69.13	73.29	89.33	107.57	113.14	118.14	124.12	128.30
100	67.33	70.06	74.22	77.93	82.36	99.33	118.50	124.34	129.56	135.81	140.17

ν = degrees of freedom.

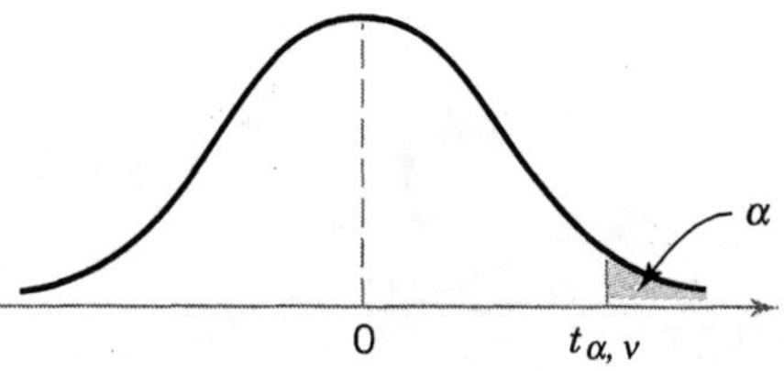

Table IV Percentage Points $t_{\alpha,\nu}$ of the t-Distribution

ν \ α	.40	.25	.10	.05	.025	.01	.005	.0025	.001	.0005
1	.325	1.000	3.078	6.314	12.706	31.821	63.657	127.32	318.31	636.62
2	.289	.816	1.886	2.920	4.303	6.965	9.925	14.089	23.326	31.598
3	.277	.765	1.638	2.353	3.182	4.541	5.841	7.453	10.213	12.924
4	.271	.741	1.533	2.132	2.776	3.747	4.604	5.598	7.173	8.610
5	.267	.727	1.476	2.015	2.571	3.365	4.032	4.773	5.893	6.869
6	.265	.718	1.440	1.943	2.447	3.143	3.707	4.317	5.208	5.959
7	.263	.711	1.415	1.895	2.365	2.998	3.499	4.029	4.785	5.408
8	.262	.706	1.397	1.860	2.306	2.896	3.355	3.833	4.501	5.041
9	.261	.703	1.383	1.833	2.262	2.821	3.250	3.690	4.297	4.781
10	.260	.700	1.372	1.812	2.228	2.764	3.169	3.581	4.144	4.587
11	.260	.697	1.363	1.796	2.201	2.718	3.106	3.497	4.025	4.437
12	.259	.695	1.356	1.782	2.179	2.681	3.055	3.428	3.930	4.318
13	.259	.694	1.350	1.771	2.160	2.650	3.012	3.372	3.852	4.221
14	.258	.692	1.345	1.761	2.145	2.624	2.977	3.326	3.787	4.140
15	.258	.691	1.341	1.753	2.131	2.602	2.947	3.286	3.733	4.073
16	.258	.690	1.337	1.746	2.120	2.583	2.921	3.252	3.686	4.015
17	.257	.689	1.333	1.740	2.110	2.567	2.898	3.222	3.646	3.965
18	.257	.688	1.330	1.734	2.101	2.552	2.878	3.197	3.610	3.922
19	.257	.688	1.328	1.729	2.093	2.539	2.861	3.174	3.579	3.883
20	.257	.687	1.325	1.725	2.086	2.528	2.845	3.153	3.552	3.850
21	.257	.686	1.323	1.721	2.080	2.518	2.831	3.135	3.527	3.819
22	.256	.686	1.321	1.717	2.074	2.508	2.819	3.119	3.505	3.792
23	.256	.685	1.319	1.714	2.069	2.500	2.807	3.104	3.485	3.767
24	.256	.685	1.318	1.711	2.064	2.492	2.797	3.091	3.467	3.745
25	.256	.684	1.316	1.708	2.060	2.485	2.787	3.078	3.450	3.725
26	.256	.684	1.315	1.706	2.056	2.479	2.779	3.067	3.435	3.707
27	.256	.684	1.314	1.703	2.052	2.473	2.771	3.057	3.421	3.690
28	.256	.683	1.313	1.701	2.048	2.467	2.763	3.047	3.408	3.674
29	.256	.683	1.311	1.699	2.045	2.462	2.756	3.038	3.396	3.659
30	.256	.683	1.310	1.697	2.042	2.457	2.750	3.030	3.385	3.646
40	.255	.681	1.303	1.684	2.021	2.423	2.704	2.971	3.307	3.551
60	.254	.679	1.296	1.671	2.000	2.390	2.660	2.915	3.232	3.460
120	.254	.677	1.289	1.658	1.980	2.358	2.617	2.860	3.160	3.373
∞	.253	.674	1.282	1.645	1.960	2.326	2.576	2.807	3.090	3.291

ν = degrees of freedom.

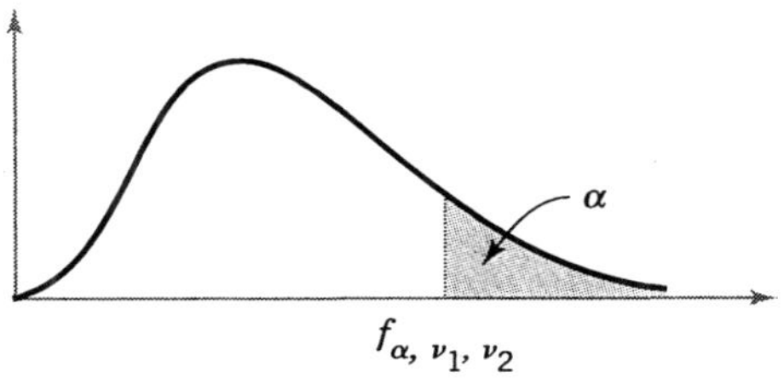

Table V Percentage Points f_{α,ν_1,ν_2} of the F-Distribution

$f_{0.25,\nu_1,\nu_2}$

Degrees of freedom for the denominator (ν_2)	Degrees of freedom for the numerator (ν_1)																		
ν_2 \ ν_1	1	2	3	4	5	6	7	8	9	10	12	15	20	24	30	40	60	120	∞
1	5.83	7.50	8.20	8.58	8.82	8.98	9.10	9.19	9.26	9.32	9.41	9.49	9.58	9.63	9.67	9.71	9.76	9.80	9.85
2	2.57	3.00	3.15	3.23	3.28	3.31	3.34	3.35	3.37	3.38	3.39	3.41	3.43	3.43	3.44	3.45	3.46	3.47	3.48
3	2.02	2.28	2.36	2.39	2.41	2.42	2.43	2.44	2.44	2.44	2.45	2.46	2.46	2.46	2.47	2.47	2.47	2.47	2.47
4	1.81	2.00	2.05	2.06	2.07	2.08	2.08	2.08	2.08	2.08	2.08	2.08	2.08	2.08	2.08	2.08	2.08	2.08	2.08
5	1.69	1.85	1.88	1.89	1.89	1.89	1.89	1.89	1.89	1.89	1.89	1.89	1.88	1.88	1.88	1.88	1.87	1.87	1.87
6	1.62	1.76	1.78	1.79	1.79	1.78	1.78	1.78	1.77	1.77	1.77	1.76	1.76	1.75	1.75	1.75	1.74	1.74	1.74
7	1.57	1.70	1.72	1.72	1.71	1.71	1.70	1.70	1.70	1.69	1.68	1.68	1.67	1.67	1.66	1.66	1.65	1.65	1.65
8	1.54	1.66	1.67	1.66	1.66	1.65	1.64	1.64	1.63	1.63	1.62	1.62	1.61	1.60	1.60	1.59	1.59	1.58	1.58
9	1.51	1.62	1.63	1.63	1.62	1.61	1.60	1.60	1.59	1.59	1.58	1.57	1.56	1.56	1.55	1.54	1.54	1.53	1.53
10	1.49	1.60	1.60	1.59	1.59	1.58	1.57	1.56	1.56	1.55	1.54	1.53	1.52	1.52	1.51	1.51	1.50	1.49	1.48
11	1.47	1.58	1.58	1.57	1.56	1.55	1.54	1.53	1.53	1.52	1.51	1.50	1.49	1.49	1.48	1.47	1.47	1.46	1.45
12	1.46	1.56	1.56	1.55	1.54	1.53	1.52	1.51	1.51	1.50	1.49	1.48	1.47	1.46	1.45	1.45	1.44	1.43	1.42
13	1.45	1.55	1.55	1.53	1.52	1.51	1.50	1.49	1.49	1.48	1.47	1.46	1.45	1.44	1.43	1.42	1.42	1.41	1.40
14	1.44	1.53	1.53	1.52	1.51	1.50	1.49	1.48	1.47	1.46	1.45	1.44	1.43	1.42	1.41	1.41	1.40	1.39	1.38
15	1.43	1.52	1.52	1.51	1.49	1.48	1.47	1.46	1.46	1.45	1.44	1.43	1.41	1.41	1.40	1.39	1.38	1.37	1.36
16	1.42	1.51	1.51	1.50	1.48	1.47	1.46	1.45	1.44	1.44	1.43	1.41	1.40	1.39	1.38	1.37	1.36	1.35	1.34
17	1.42	1.51	1.50	1.49	1.47	1.46	1.45	1.44	1.43	1.43	1.41	1.40	1.39	1.38	1.37	1.36	1.35	1.34	1.33
18	1.41	1.50	1.49	1.48	1.46	1.45	1.44	1.43	1.42	1.42	1.40	1.39	1.38	1.37	1.36	1.35	1.34	1.33	1.32
19	1.41	1.49	1.49	1.47	1.46	1.44	1.43	1.42	1.41	1.41	1.40	1.38	1.37	1.36	1.35	1.34	1.33	1.32	1.30
20	1.40	1.49	1.48	1.47	1.45	1.44	1.43	1.42	1.41	1.40	1.39	1.37	1.36	1.35	1.34	1.33	1.32	1.31	1.29
21	1.40	1.48	1.48	1.46	1.44	1.43	1.42	1.41	1.40	1.39	1.38	1.37	1.35	1.34	1.33	1.32	1.31	1.30	1.28
22	1.40	1.48	1.47	1.45	1.44	1.42	1.41	1.40	1.39	1.39	1.37	1.36	1.34	1.33	1.32	1.31	1.30	1.29	1.28
23	1.39	1.47	1.47	1.45	1.43	1.42	1.41	1.40	1.39	1.38	1.37	1.35	1.34	1.33	1.32	1.31	1.30	1.28	1.27
24	1.39	1.47	1.46	1.44	1.43	1.41	1.40	1.39	1.38	1.38	1.36	1.35	1.33	1.32	1.31	1.30	1.29	1.28	1.26
25	1.39	1.47	1.46	1.44	1.42	1.41	1.40	1.39	1.38	1.37	1.36	1.34	1.33	1.32	1.31	1.29	1.28	1.27	1.25
26	1.38	1.46	1.45	1.44	1.42	1.41	1.39	1.38	1.37	1.37	1.35	1.34	1.32	1.31	1.30	1.29	1.28	1.26	1.25
27	1.38	1.46	1.45	1.43	1.42	1.40	1.39	1.38	1.37	1.36	1.35	1.33	1.32	1.31	1.30	1.28	1.27	1.26	1.24
28	1.38	1.46	1.45	1.43	1.41	1.40	1.39	1.38	1.37	1.36	1.34	1.33	1.31	1.30	1.29	1.28	1.27	1.25	1.24
29	1.38	1.45	1.45	1.43	1.41	1.40	1.38	1.37	1.36	1.35	1.34	1.32	1.31	1.30	1.29	1.27	1.26	1.25	1.23
30	1.38	1.45	1.44	1.42	1.41	1.39	1.38	1.37	1.36	1.35	1.34	1.32	1.30	1.29	1.28	1.27	1.26	1.24	1.23
40	1.36	1.44	1.42	1.40	1.39	1.37	1.36	1.35	1.34	1.33	1.31	1.30	1.28	1.26	1.25	1.24	1.22	1.21	1.19
60	1.35	1.42	1.41	1.38	1.37	1.35	1.33	1.32	1.31	1.30	1.29	1.27	1.25	1.24	1.22	1.21	1.19	1.17	1.15
120	1.34	1.40	1.39	1.37	1.35	1.33	1.31	1.30	1.29	1.28	1.26	1.24	1.22	1.21	1.19	1.18	1.16	1.13	1.10
∞	1.32	1.39	1.37	1.35	1.33	1.31	1.29	1.28	1.27	1.25	1.24	1.22	1.19	1.18	1.16	1.14	1.12	1.08	1.00

Table V Percentage Points of the *F*-Distribution (*continued*)

$f_{0.10,\nu_1,\nu_2}$

Degrees of freedom for the numerator (ν_1); Degrees of freedom for the denominator (ν_2)

ν_2 \ ν_1	1	2	3	4	5	6	7	8	9	10	12	15	20	24	30	40	60	120	∞
1	39.86	49.50	53.59	55.83	57.24	58.20	58.91	59.44	59.86	60.19	60.71	61.22	61.74	62.00	62.26	62.53	62.79	63.06	63.33
2	8.53	9.00	9.16	9.24	9.29	9.33	9.35	9.37	9.38	9.39	9.41	9.42	9.44	9.45	9.46	9.47	9.47	9.48	9.49
3	5.54	5.46	5.39	5.34	5.31	5.28	5.27	5.25	5.24	5.23	5.22	5.20	5.18	5.18	5.17	5.16	5.15	5.14	5.13
4	4.54	4.32	4.19	4.11	4.05	4.01	3.98	3.95	3.94	3.92	3.90	3.87	3.84	3.83	3.82	3.80	3.79	3.78	3.76
5	4.06	3.78	3.62	3.52	3.45	3.40	3.37	3.34	3.32	3.30	3.27	3.24	3.21	3.19	3.17	3.16	3.14	3.12	3.10
6	3.78	3.46	3.29	3.18	3.11	3.05	3.01	2.98	2.96	2.94	2.90	2.87	2.84	2.82	2.80	2.78	2.76	2.74	2.72
7	3.59	3.26	3.07	2.96	2.88	2.83	2.78	2.75	2.72	2.70	2.67	2.63	2.59	2.58	2.56	2.54	2.51	2.49	2.47
8	3.46	3.11	2.92	2.81	2.73	2.67	2.62	2.59	2.56	2.54	2.50	2.46	2.42	2.40	2.38	2.36	2.34	2.32	2.29
9	3.36	3.01	2.81	2.69	2.61	2.55	2.51	2.47	2.44	2.42	2.38	2.34	2.30	2.28	2.25	2.23	2.21	2.18	2.16
10	3.29	2.92	2.73	2.61	2.52	2.46	2.41	2.38	2.35	2.32	2.28	2.24	2.20	2.18	2.16	2.13	2.11	2.08	2.06
11	3.23	2.86	2.66	2.54	2.45	2.39	2.34	2.30	2.27	2.25	2.21	2.17	2.12	2.10	2.08	2.05	2.03	2.00	1.97
12	3.18	2.81	2.61	2.48	2.39	2.33	2.28	2.24	2.21	2.19	2.15	2.10	2.06	2.04	2.01	1.99	1.96	1.93	1.90
13	3.14	2.76	2.56	2.43	2.35	2.28	2.23	2.20	2.16	2.14	2.10	2.05	2.01	1.98	1.96	1.93	1.90	1.88	1.85
14	3.10	2.73	2.52	2.39	2.31	2.24	2.19	2.15	2.12	2.10	2.05	2.01	1.96	1.94	1.91	1.89	1.86	1.83	1.80
15	3.07	2.70	2.49	2.36	2.27	2.21	2.16	2.12	2.09	2.06	2.02	1.97	1.92	1.90	1.87	1.85	1.82	1.79	1.76
16	3.05	2.67	2.46	2.33	2.24	2.18	2.13	2.09	2.06	2.03	1.99	1.94	1.89	1.87	1.84	1.81	1.78	1.75	1.72
17	3.03	2.64	2.44	2.31	2.22	2.15	2.10	2.06	2.03	2.00	1.96	1.91	1.86	1.84	1.81	1.78	1.75	1.72	1.69
18	3.01	2.62	2.42	2.29	2.20	2.13	2.08	2.04	2.00	1.98	1.93	1.89	1.84	1.81	1.78	1.75	1.72	1.69	1.66
19	2.99	2.61	2.40	2.27	2.18	2.11	2.06	2.02	1.98	1.96	1.91	1.86	1.81	1.79	1.76	1.73	1.70	1.67	1.63
20	2.97	2.59	2.38	2.25	2.16	2.09	2.04	2.00	1.96	1.94	1.89	1.84	1.79	1.77	1.74	1.71	1.68	1.64	1.61
21	2.96	2.57	2.36	2.23	2.14	2.08	2.02	1.98	1.95	1.92	1.87	1.83	1.78	1.75	1.72	1.69	1.66	1.62	1.59
22	2.95	2.56	2.35	2.22	2.13	2.06	2.01	1.97	1.93	1.90	1.86	1.81	1.76	1.73	1.70	1.67	1.64	1.60	1.57
23	2.94	2.55	2.34	2.21	2.11	2.05	1.99	1.95	1.92	1.89	1.84	1.80	1.74	1.72	1.69	1.66	1.62	1.59	1.55
24	2.93	2.54	2.33	2.19	2.10	2.04	1.98	1.94	1.91	1.88	1.83	1.78	1.73	1.70	1.67	1.64	1.61	1.57	1.53
25	2.92	2.53	2.32	2.18	2.09	2.02	1.97	1.93	1.89	1.87	1.82	1.77	1.72	1.69	1.66	1.63	1.59	1.56	1.52
26	2.91	2.52	2.31	2.17	2.08	2.01	1.96	1.92	1.88	1.86	1.81	1.76	1.71	1.68	1.65	1.61	1.58	1.54	1.50
27	2.90	2.51	2.30	2.17	2.07	2.00	1.95	1.91	1.87	1.85	1.80	1.75	1.70	1.67	1.64	1.60	1.57	1.53	1.49
28	2.89	2.50	2.29	2.16	2.06	2.00	1.94	1.90	1.87	1.84	1.79	1.74	1.69	1.66	1.63	1.59	1.56	1.52	1.48
29	2.89	2.50	2.28	2.15	2.06	1.99	1.93	1.89	1.86	1.83	1.78	1.73	1.68	1.65	1.62	1.58	1.55	1.51	1.47
30	2.88	2.49	2.28	2.14	2.03	1.98	1.93	1.88	1.85	1.82	1.77	1.72	1.67	1.64	1.61	1.57	1.54	1.50	1.46
40	2.84	2.44	2.23	2.09	2.00	1.93	1.87	1.83	1.79	1.76	1.71	1.66	1.61	1.57	1.54	1.51	1.47	1.42	1.38
60	2.79	2.39	2.18	2.04	1.95	1.87	1.82	1.77	1.74	1.71	1.66	1.60	1.54	1.51	1.48	1.44	1.40	1.35	1.29
120	2.75	2.35	2.13	1.99	1.90	1.82	1.77	1.72	1.68	1.65	1.60	1.55	1.48	1.45	1.41	1.37	1.32	1.26	1.19
∞	2.71	2.30	2.08	1.94	1.85	1.77	1.72	1.67	1.63	1.60	1.55	1.49	1.42	1.38	1.34	1.30	1.24	1.17	1.00

Table V Percentage Points of the F-Distribution (*continued*)

$$f_{0.05,\nu_1,\nu_2}$$

Degrees of freedom for the denominator (ν_2); Degrees of freedom for the numerator (ν_1)

ν_2 \ ν_1	1	2	3	4	5	6	7	8	9	10	12	15	20	24	30	40	60	120	∞
1	161.4	199.5	215.7	224.6	230.2	234.0	236.8	238.9	240.5	241.9	243.9	245.9	248.0	249.1	250.1	251.1	252.2	253.3	254.3
2	18.51	19.00	19.16	19.25	19.30	19.33	19.35	19.37	19.38	19.40	19.41	19.43	19.45	19.45	19.46	19.47	19.48	19.49	19.50
3	10.13	9.55	9.28	9.12	9.01	8.94	8.89	8.85	8.81	8.79	8.74	8.70	8.66	8.64	8.62	8.59	8.57	8.55	8.53
4	7.71	6.94	6.59	6.39	6.26	6.16	6.09	6.04	6.00	5.96	5.91	5.86	5.80	5.77	5.75	5.72	5.69	5.66	5.63
5	6.61	5.79	5.41	5.19	5.05	4.95	4.88	4.82	4.77	4.74	4.68	4.62	4.56	4.53	4.50	4.46	4.43	4.40	4.36
6	5.99	5.14	4.76	4.53	4.39	4.28	4.21	4.15	4.10	4.06	4.00	3.94	3.87	3.84	3.81	3.77	3.74	3.70	3.67
7	5.59	4.74	4.35	4.12	3.97	3.87	3.79	3.73	3.68	3.64	3.57	3.51	3.44	3.41	3.38	3.34	3.30	3.27	3.23
8	5.32	4.46	4.07	3.84	3.69	3.58	3.50	3.44	3.39	3.35	3.28	3.22	3.15	3.12	3.08	3.04	3.01	2.97	2.93
9	5.12	4.26	3.86	3.63	3.48	3.37	3.29	3.23	3.18	3.14	3.07	3.01	2.94	2.90	2.86	2.83	2.79	2.75	2.71
10	4.96	4.10	3.71	3.48	3.33	3.22	3.14	3.07	3.02	2.98	2.91	2.85	2.77	2.74	2.70	2.66	2.62	2.58	2.54
11	4.84	3.98	3.59	3.36	3.20	3.09	3.01	2.95	2.90	2.85	2.79	2.72	2.65	2.61	2.57	2.53	2.49	2.45	2.40
12	4.75	3.89	3.49	3.26	3.11	3.00	2.91	2.85	2.80	2.75	2.69	2.62	2.54	2.51	2.47	2.43	2.38	2.34	2.30
13	4.67	3.81	3.41	3.18	3.03	2.92	2.83	2.77	2.71	2.67	2.60	2.53	2.46	2.42	2.38	2.34	2.30	2.25	2.21
14	4.60	3.74	3.34	3.11	2.96	2.85	2.76	2.70	2.65	2.60	2.53	2.46	2.39	2.35	2.31	2.27	2.22	2.18	2.13
15	4.54	3.68	3.29	3.06	2.90	2.79	2.71	2.64	2.59	2.54	2.48	2.40	2.33	2.29	2.25	2.20	2.16	2.11	2.07
16	4.49	3.63	3.24	3.01	2.85	2.74	2.66	2.59	2.54	2.49	2.42	2.35	2.28	2.24	2.19	2.15	2.11	2.06	2.01
17	4.45	3.59	3.20	2.96	2.81	2.70	2.61	2.55	2.49	2.45	2.38	2.31	2.23	2.19	2.15	2.10	2.06	2.01	1.96
18	4.41	3.55	3.16	2.93	2.77	2.66	2.58	2.51	2.46	2.41	2.34	2.27	2.19	2.15	2.11	2.06	2.02	1.97	1.92
19	4.38	3.52	3.13	2.90	2.74	2.63	2.54	2.48	2.42	2.38	2.31	2.23	2.16	2.11	2.07	2.03	1.98	1.93	1.88
20	4.35	3.49	3.10	2.87	2.71	2.60	2.51	2.45	2.39	2.35	2.28	2.20	2.12	2.08	2.04	1.99	1.95	1.90	1.84
21	4.32	3.47	3.07	2.84	2.68	2.57	2.49	2.42	2.37	2.32	2.25	2.18	2.10	2.05	2.01	1.96	1.92	1.87	1.81
22	4.30	3.44	3.05	2.82	2.66	2.55	2.46	2.40	2.34	2.30	2.23	2.15	2.07	2.03	1.98	1.94	1.89	1.84	1.78
23	4.28	3.42	3.03	2.80	2.64	2.53	2.44	2.37	2.32	2.27	2.20	2.13	2.05	2.01	1.96	1.91	1.86	1.81	1.76
24	4.26	3.40	3.01	2.78	2.62	2.51	2.42	2.36	2.30	2.25	2.18	2.11	2.03	1.98	1.94	1.89	1.84	1.79	1.73
25	4.24	3.39	2.99	2.76	2.60	2.49	2.40	2.34	2.28	2.24	2.16	2.09	2.01	1.96	1.92	1.87	1.82	1.77	1.71
26	4.23	3.37	2.98	2.74	2.59	2.47	2.39	2.32	2.27	2.22	2.15	2.07	1.99	1.95	1.90	1.85	1.80	1.75	1.69
27	4.21	3.35	2.96	2.73	2.57	2.46	2.37	2.31	2.25	2.20	2.13	2.06	1.97	1.93	1.88	1.84	1.79	1.73	1.67
28	4.20	3.34	2.95	2.71	2.56	2.45	2.36	2.29	2.24	2.19	2.12	2.04	1.96	1.91	1.87	1.82	1.77	1.71	1.65
29	4.18	3.33	2.93	2.70	2.55	2.43	2.35	2.28	2.22	2.18	2.10	2.03	1.94	1.90	1.85	1.81	1.75	1.70	1.64
30	4.17	3.32	2.92	2.69	2.53	2.42	2.33	2.27	2.21	2.16	2.09	2.01	1.93	1.89	1.84	1.79	1.74	1.68	1.62
40	4.08	3.23	2.84	2.61	2.45	2.34	2.25	2.18	2.12	2.08	2.00	1.92	1.84	1.79	1.74	1.69	1.64	1.58	1.51
60	4.00	3.15	2.76	2.53	2.37	2.25	2.17	2.10	2.04	1.99	1.92	1.84	1.75	1.70	1.65	1.59	1.53	1.47	1.39
120	3.92	3.07	2.68	2.45	2.29	2.17	2.09	2.02	1.96	1.91	1.83	1.75	1.66	1.61	1.55	1.55	1.43	1.35	1.25
∞	3.84	3.00	2.60	2.37	2.21	2.10	2.01	1.94	1.88	1.83	1.75	1.67	1.57	1.52	1.46	1.39	1.32	1.22	1.00

Table V Percentage Points of the *F*-Distribution (*continued*)

$f_{0.025,\nu_1,\nu_2}$

Degrees of freedom for the numerator (ν_1); Degrees of freedom for the denominator (ν_2)

ν_2 \ ν_1	1	2	3	4	5	6	7	8	9	10	12	15	20	24	30	40	60	120	∞
1	647.8	799.5	864.2	899.6	921.8	937.1	948.2	956.7	963.3	968.6	976.7	984.9	993.1	997.2	1001	1006	1010	1014	1018
2	38.51	39.00	39.17	39.25	39.30	39.33	39.36	39.37	39.39	39.40	39.41	39.43	39.45	39.46	39.46	39.47	39.48	39.49	39.50
3	17.44	16.04	15.44	15.10	14.88	14.73	14.62	14.54	14.47	14.42	14.34	14.25	14.17	14.12	14.08	14.04	13.99	13.95	13.90
4	12.22	10.65	9.98	9.60	9.36	9.20	9.07	8.98	8.90	8.84	8.75	8.66	8.56	8.51	8.46	8.41	8.36	8.31	8.26
5	10.01	8.43	7.76	7.39	7.15	6.98	6.85	6.76	6.68	6.62	6.52	6.43	6.33	6.28	6.23	6.18	6.12	6.07	6.02
6	8.81	7.26	6.60	6.23	5.99	5.82	5.70	5.60	5.52	5.46	5.37	5.27	5.17	5.12	5.07	5.01	4.96	4.90	4.85
7	8.07	6.54	5.89	5.52	5.29	5.12	4.99	4.90	4.82	4.76	4.67	4.57	4.47	4.42	4.36	4.31	4.25	4.20	4.14
8	7.57	6.06	5.42	5.05	4.82	4.65	4.53	4.43	4.36	4.30	4.20	4.10	4.00	3.95	3.89	3.84	3.78	3.73	3.67
9	7.21	5.71	5.08	4.72	4.48	4.32	4.20	4.10	4.03	3.96	3.87	3.77	3.67	3.61	3.56	3.51	3.45	3.39	3.33
10	6.94	5.46	4.83	4.47	4.24	4.07	3.95	3.85	3.78	3.72	3.62	3.52	3.42	3.37	3.31	3.26	3.20	3.14	3.08
11	6.72	5.26	4.63	4.28	4.04	3.88	3.76	3.66	3.59	3.53	3.43	3.33	3.23	3.17	3.12	3.06	3.00	2.94	2.88
12	6.55	5.10	4.47	4.12	3.89	3.73	3.61	3.51	3.44	3.37	3.28	3.18	3.07	3.02	2.96	2.91	2.85	2.79	2.72
13	6.41	4.97	4.35	4.00	3.77	3.60	3.48	3.39	3.31	3.25	3.15	3.05	2.95	2.89	2.84	2.78	2.72	2.66	2.60
14	6.30	4.86	4.24	3.89	3.66	3.50	3.38	3.29	3.21	3.15	3.05	2.95	2.84	2.79	2.73	2.67	2.61	2.55	2.49
15	6.20	4.77	4.15	3.80	3.58	3.41	3.29	3.20	3.12	3.06	2.96	2.86	2.76	2.70	2.64	2.59	2.52	2.46	2.40
16	6.12	4.69	4.08	3.73	3.50	3.34	3.22	3.12	3.05	2.99	2.89	2.79	2.68	2.63	2.57	2.51	2.45	2.38	2.32
17	6.04	4.62	4.01	3.66	3.44	3.28	3.16	3.06	2.98	2.92	2.82	2.72	2.62	2.56	2.50	2.44	2.38	2.32	2.25
18	5.98	4.56	3.95	3.61	3.38	3.22	3.10	3.01	2.93	2.87	2.77	2.67	2.56	2.50	2.44	2.38	2.32	2.26	2.19
19	5.92	4.51	3.90	3.56	3.33	3.17	3.05	2.96	2.88	2.82	2.72	2.62	2.51	2.45	2.39	2.33	2.27	2.20	2.13
20	5.87	4.46	3.86	3.51	3.29	3.13	3.01	2.91	2.84	2.77	2.68	2.57	2.46	2.41	2.35	2.29	2.22	2.16	2.09
21	5.83	4.42	3.82	3.48	3.25	3.09	2.97	2.87	2.80	2.73	2.64	2.53	2.42	2.37	2.31	2.25	2.18	2.11	2.04
22	5.79	4.38	3.78	3.44	3.22	3.05	2.93	2.84	2.76	2.70	2.60	2.50	2.39	2.33	2.27	2.21	2.14	2.08	2.00
23	5.75	4.35	3.75	3.41	3.18	3.02	2.90	2.81	2.73	2.67	2.57	2.47	2.36	2.30	2.24	2.18	2.11	2.04	1.97
24	5.72	4.32	3.72	3.38	3.15	2.99	2.87	2.78	2.70	2.64	2.54	2.44	2.33	2.27	2.21	2.15	2.08	2.01	1.94
25	5.69	4.29	3.69	3.35	3.13	2.97	2.85	2.75	2.68	2.61	2.51	2.41	2.30	2.24	2.18	2.12	2.05	1.98	1.91
26	5.66	4.27	3.67	3.33	3.10	2.94	2.82	2.73	2.65	2.59	2.49	2.39	2.28	2.22	2.16	2.09	2.03	1.95	1.88
27	5.63	4.24	3.65	3.31	3.08	2.92	2.80	2.71	2.63	2.57	2.47	2.36	2.25	2.19	2.13	2.07	2.00	1.93	1.85
28	5.61	4.22	3.63	3.29	3.06	2.90	2.78	2.69	2.61	2.55	2.45	2.34	2.23	2.17	2.11	2.05	1.98	1.91	1.83
29	5.59	4.20	3.61	3.27	3.04	2.88	2.76	2.67	2.59	2.53	2.43	2.32	2.21	2.15	2.09	2.03	1.96	1.89	1.81
30	5.57	4.18	3.59	3.25	3.03	2.87	2.75	2.65	2.57	2.51	2.41	2.31	2.20	2.14	2.07	2.01	1.94	1.87	1.79
40	5.42	4.05	3.46	3.13	2.90	2.74	2.62	2.53	2.45	2.39	2.29	2.18	2.07	2.01	1.94	1.88	1.80	1.72	1.64
60	5.29	3.93	3.34	3.01	2.79	2.63	2.51	2.41	2.33	2.27	2.17	2.06	1.94	1.88	1.82	1.74	1.67	1.58	1.48
120	5.15	3.80	3.23	2.89	2.67	2.52	2.39	2.30	2.22	2.16	2.05	1.94	1.82	1.76	1.69	1.61	1.53	1.43	1.31
∞	5.02	3.69	3.12	2.79	2.57	2.41	2.29	2.19	2.11	2.05	1.94	1.83	1.71	1.64	1.57	1.48	1.39	1.27	1.00

Table V Percentage Points of the F-Distribution (*continued*)

$$f_{0.01,\nu_1,\nu_2}$$

ν_2 \ ν_1	Degrees of freedom for the numerator (ν_1) 1	2	3	4	5	6	7	8	9	10	12	15	20	24	30	40	60	120	∞
1	4052	4999.5	5403	5625	5764	5859	5928	5982	6022	6056	6106	6157	6209	6235	6261	6287	6313	6339	6366
2	98.50	99.00	99.17	99.25	99.30	99.33	99.36	99.37	99.39	99.40	99.42	99.43	99.45	99.46	99.47	99.47	99.48	99.49	99.50
3	34.12	30.82	29.46	28.71	28.24	27.91	27.67	27.49	27.35	27.23	27.05	26.87	26.69	26.00	26.50	26.41	26.32	26.22	26.13
4	21.20	18.00	16.69	15.98	15.52	15.21	14.98	14.80	14.66	14.55	14.37	14.20	14.02	13.93	13.84	13.75	13.65	13.56	13.46
5	16.26	13.27	12.06	11.39	10.97	10.67	10.46	10.29	10.16	10.05	9.89	9.72	9.55	9.47	9.38	9.29	9.20	9.11	9.02
6	13.75	10.92	9.78	9.15	8.75	8.47	8.26	8.10	7.98	7.87	7.72	7.56	7.40	7.31	7.23	7.14	7.06	6.97	6.88
7	12.25	9.55	8.45	7.85	7.46	7.19	6.99	6.84	6.72	6.62	6.47	6.31	6.16	6.07	5.99	5.91	5.82	5.74	5.65
8	11.26	8.65	7.59	7.01	6.63	6.37	6.18	6.03	5.91	5.81	5.67	5.52	5.36	5.28	5.20	5.12	5.03	4.95	4.46
9	10.56	8.02	6.99	6.42	6.06	5.80	5.61	5.47	5.35	5.26	5.11	4.96	4.81	4.73	4.65	4.57	4.48	4.40	4.31
10	10.04	7.56	6.55	5.99	5.64	5.39	5.20	5.06	4.94	4.85	4.71	4.56	4.41	4.33	4.25	4.17	4.08	4.00	3.91
11	9.65	7.21	6.22	5.67	5.32	5.07	4.89	4.74	4.63	4.54	4.40	4.25	4.10	4.02	3.94	3.86	3.78	3.69	3.60
12	9.33	6.93	5.95	5.41	5.06	4.82	4.64	4.50	4.39	4.30	4.16	4.01	3.86	3.78	3.70	3.62	3.54	3.45	3.36
13	9.07	6.70	5.74	5.21	4.86	4.62	4.44	4.30	4.19	4.10	3.96	3.82	3.66	3.59	3.51	3.43	3.34	3.25	3.17
14	8.86	6.51	5.56	5.04	4.69	4.46	4.28	4.14	4.03	3.94	3.80	3.66	3.51	3.43	3.35	3.27	3.18	3.09	3.00
15	8.68	6.36	5.42	4.89	4.36	4.32	4.14	4.00	3.89	3.80	3.67	3.52	3.37	3.29	3.21	3.13	3.05	2.96	2.87
16	8.53	6.23	5.29	4.77	4.44	4.20	4.03	3.89	3.78	3.69	3.55	3.41	3.26	3.18	3.10	3.02	2.93	2.84	2.75
17	8.40	6.11	5.18	4.67	4.34	4.10	3.93	3.79	3.68	3.59	3.46	3.31	3.16	3.08	3.00	2.92	2.83	2.75	2.65
18	8.29	6.01	5.09	4.58	4.25	4.01	3.84	3.71	3.60	3.51	3.37	3.23	3.08	3.00	2.92	2.84	2.75	2.66	2.57
19	8.18	5.93	5.01	4.50	4.17	3.94	3.77	3.63	3.52	3.43	3.30	3.15	3.00	2.92	2.84	2.76	2.67	2.58	2.59
20	8.10	5.85	4.94	4.43	4.10	3.87	3.70	3.56	3.46	3.37	3.23	3.09	2.94	2.86	2.78	2.69	2.61	2.52	2.42
21	8.02	5.78	4.87	4.37	4.04	3.81	3.64	3.51	3.40	3.31	3.17	3.03	2.88	2.80	2.72	2.64	2.55	2.46	2.36
22	7.95	5.72	4.82	4.31	3.99	3.76	3.59	3.45	3.35	3.26	3.12	2.98	2.83	2.75	2.67	2.58	2.50	2.40	2.31
23	7.88	5.66	4.76	4.26	3.94	3.71	3.54	3.41	3.30	3.21	3.07	2.93	2.78	2.70	2.62	2.54	2.45	2.35	2.26
24	7.82	5.61	4.72	4.22	3.90	3.67	3.50	3.36	3.26	3.17	3.03	2.89	2.74	2.66	2.58	2.49	2.40	2.31	2.21
25	7.77	5.57	4.68	4.18	3.85	3.63	3.46	3.32	3.22	3.13	2.99	2.85	2.70	2.62	2.54	2.45	2.36	2.27	2.17
26	7.72	5.53	4.64	4.14	3.82	3.59	3.42	3.29	3.18	3.09	2.96	2.81	2.66	2.58	2.50	2.42	2.33	2.23	2.13
27	7.68	5.49	4.60	4.11	3.78	3.56	3.39	3.26	3.15	3.06	2.93	2.78	2.63	2.55	2.47	2.38	2.29	2.20	2.10
28	7.64	5.45	4.57	4.07	3.75	3.53	3.36	3.23	3.12	3.03	2.90	2.75	2.60	2.52	2.44	2.35	2.26	2.17	2.06
29	7.60	5.42	4.54	4.04	3.73	3.50	3.33	3.20	3.09	3.00	2.87	2.73	2.57	2.49	2.41	2.33	2.23	2.14	2.03
30	7.56	5.39	4.51	4.02	3.70	3.47	3.30	3.17	3.07	2.98	2.84	2.70	2.55	2.47	2.39	2.30	2.21	2.11	2.01
40	7.31	5.18	4.31	3.83	3.51	3.29	3.12	2.99	2.89	2.80	2.66	2.52	2.37	2.29	2.20	2.11	2.02	1.92	1.80
60	7.08	4.98	4.13	3.65	3.34	3.12	2.95	2.82	2.72	2.63	2.50	2.35	2.20	2.12	2.03	1.94	1.84	1.73	1.60
120	6.85	4.79	3.95	3.48	3.17	2.96	2.79	2.66	2.56	2.47	2.34	2.19	2.03	1.95	1.86	1.76	1.66	1.53	1.38
∞	6.63	4.61	3.78	3.32	3.02	2.80	2.64	2.51	2.41	2.32	2.18	2.04	1.88	1.79	1.70	1.59	1.47	1.32	1.00

Degrees of freedom for the denominator (ν_2)

Chart VI Operating Characteristic Curves

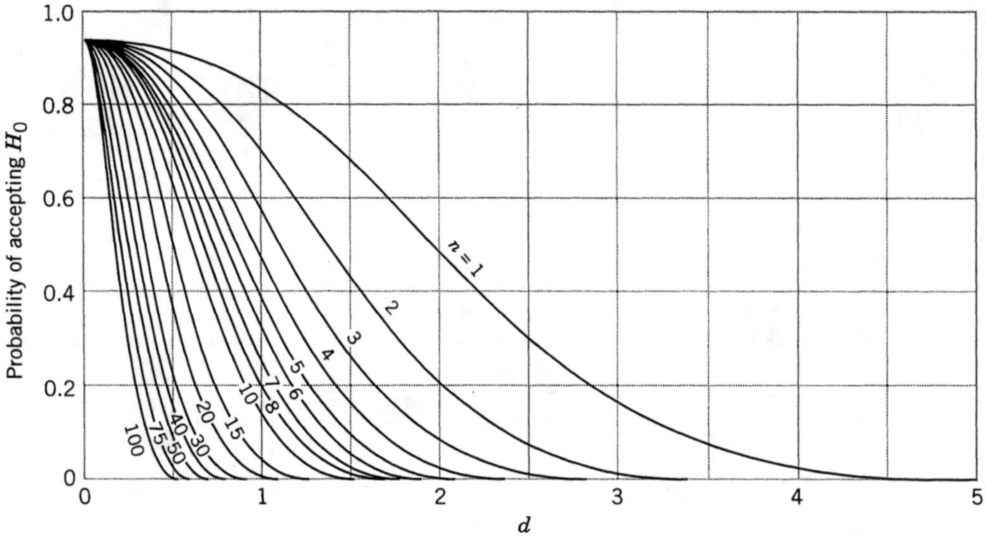

(*a*) *OC* curves for different values of *n* for the two-sided normal test for a level of significance α = 0.05.

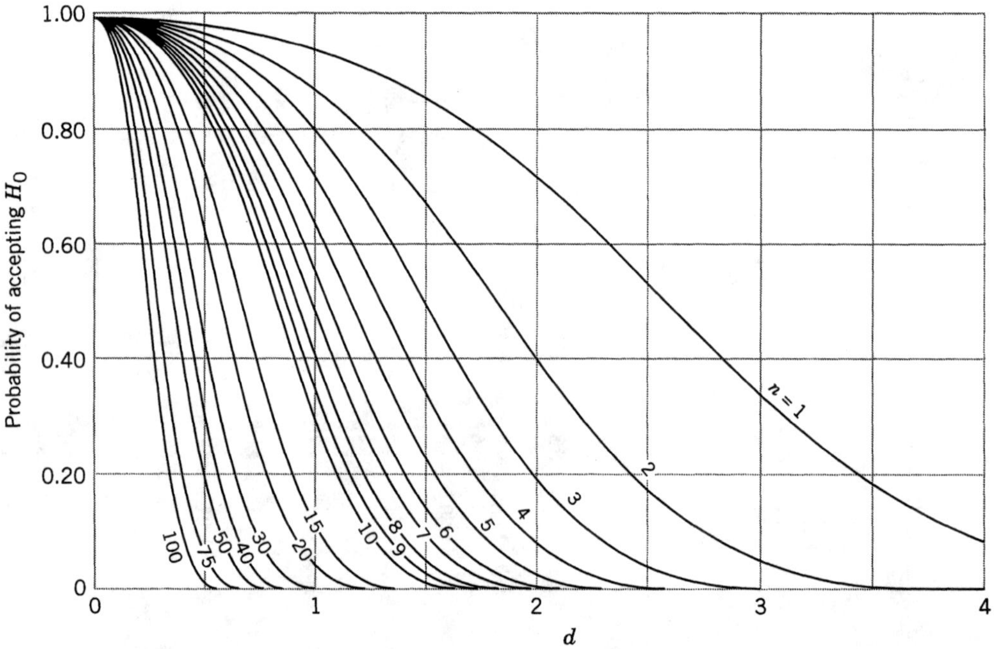

(*b*) *OC* curves for different values of *n* for the two-sided normal test for a level of significance α = 0.01.

Source: Charts VI*a, e, f, k, m,* and *q* are reproduced with permission from "Operating Characteristics for the Common Statistical Tests of Significance," by C. L. Ferris, F. E. Grubbs, and C. L. Weaver, *Annals of Mathematical Statistics,* June 1946.
Charts VI*b, c, d, g, h, i, j, l, n, o, p,* and *r* are reproduced with permission from *Engineering Statistics,* 2nd Edition, by A. H. Bowker and G. J. Lieberman, Prentice-Hall, 1972.

Chart VI Operating Characteristic Curves (*continued*)

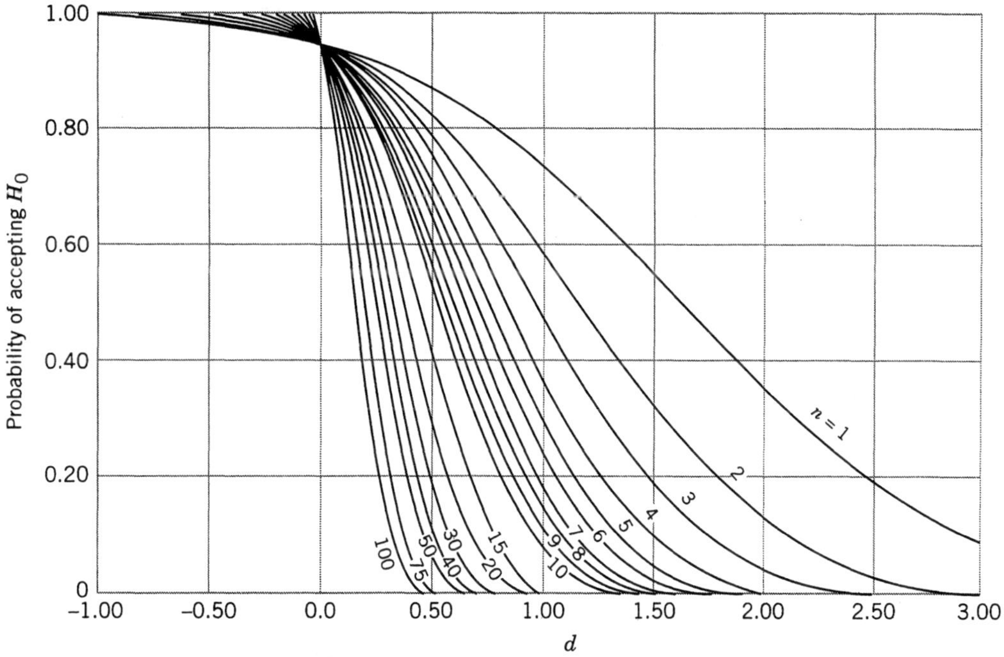

(*c*) *OC* curves for different values of *n* for the one-sided normal test for a level of significance $\alpha = 0.05$.

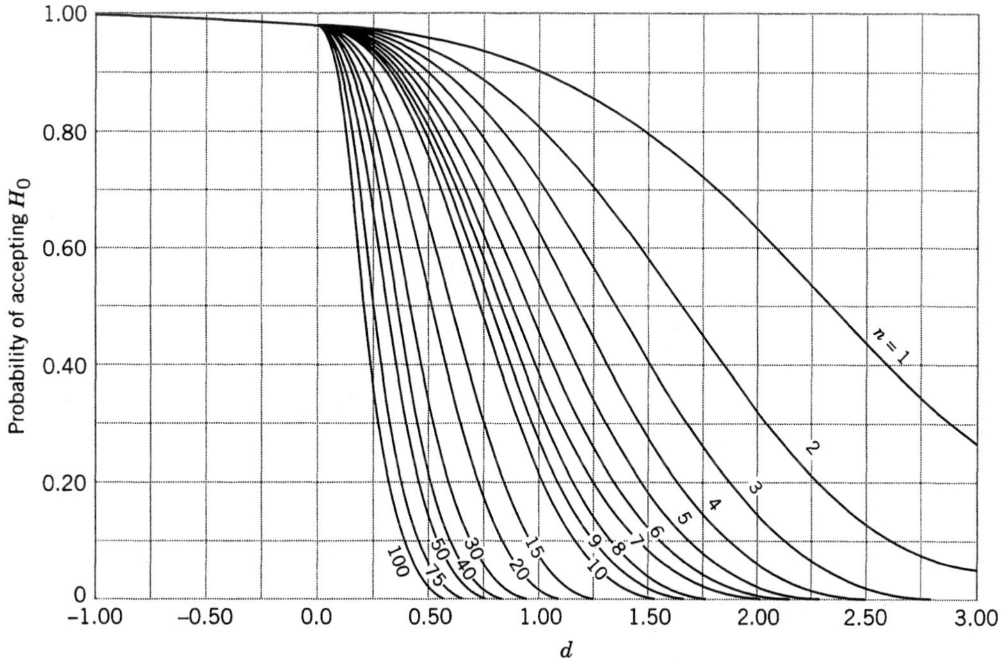

(*d*) *OC* curves for different values of *n* for the one-sided normal test for a level of significance $\alpha = 0.01$.

Chart VI Operating Characteristic Curves (*continued*)

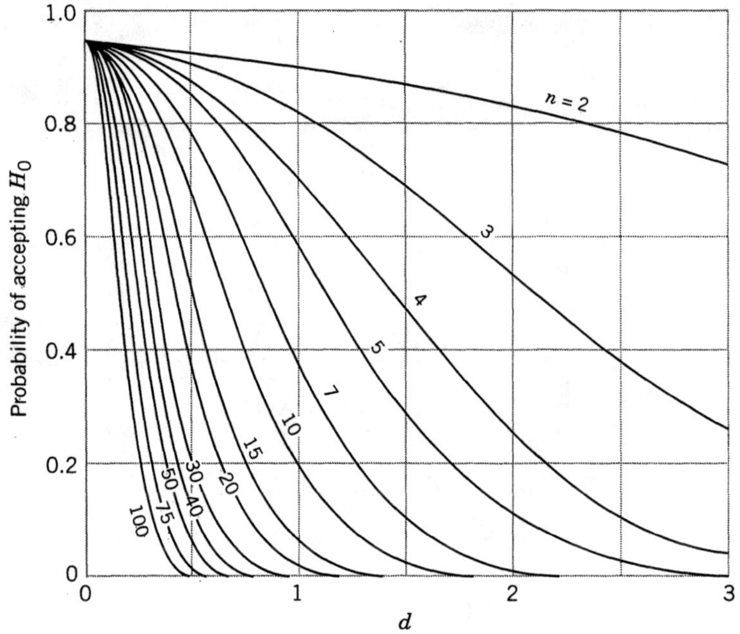

(*e*) *OC* curves for different values of *n* for the two-sided *t*-test for a level of significance $\alpha = 0.05$

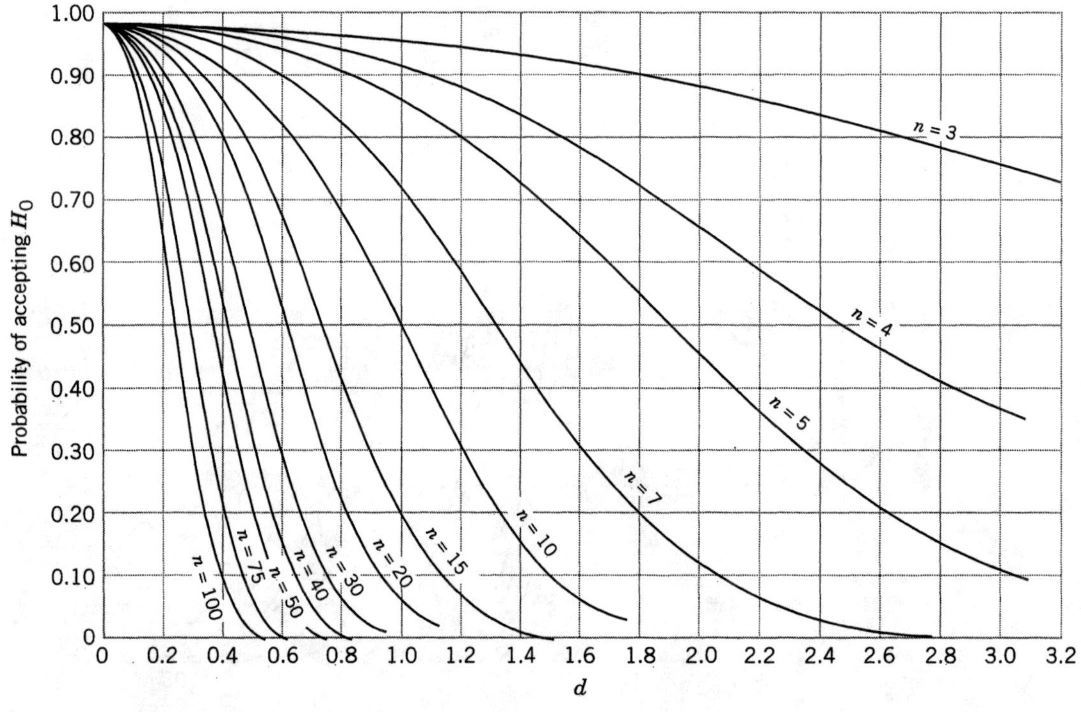

(*f*) *OC* curves for different values of *n* for the two-sided *t*-test for a level of significance $\alpha = 0.01$.

Chart VI Operating Characteristic Curves (*continued*)

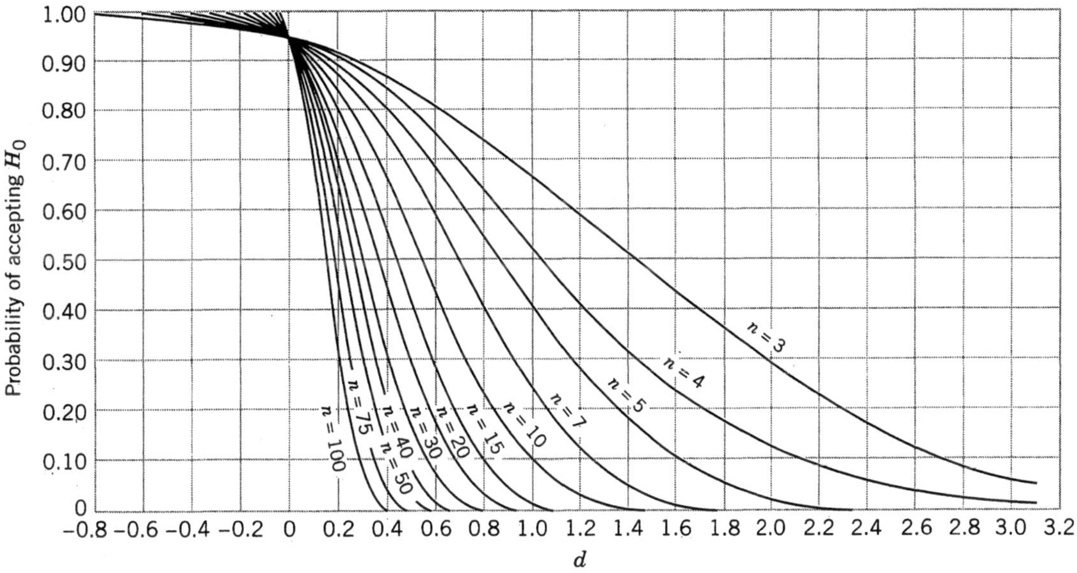

(*g*) *OC* curves for different values of n for the one-sided t-test for a level of significance $\alpha = 0.05$.

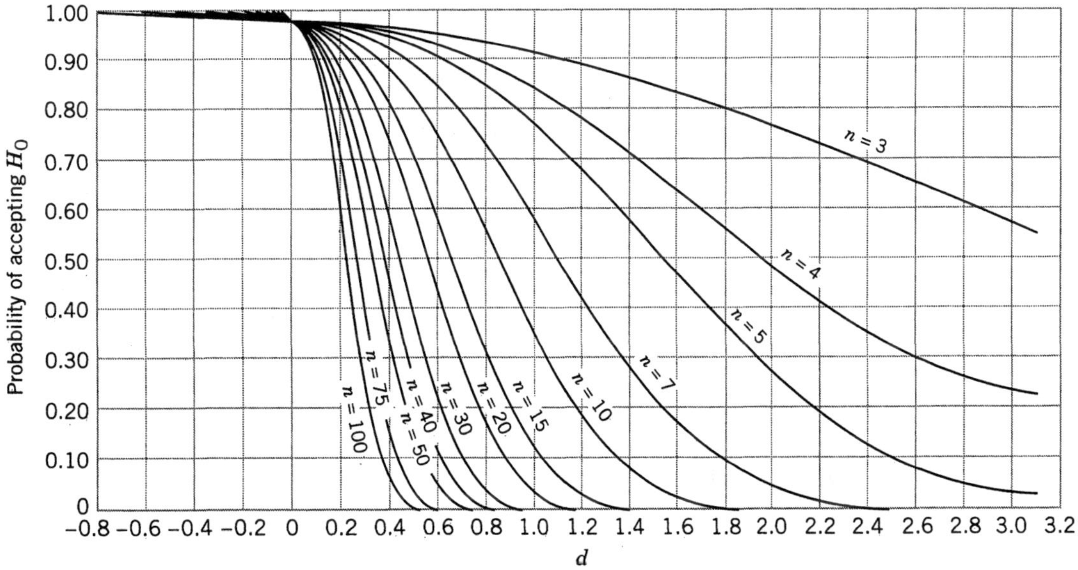

(*h*) *OC* curves for different values of n for the one-sided t-test for a level of significance $\alpha = 0.01$.

Chart VI Operating Characteristic Curves (*continued*)

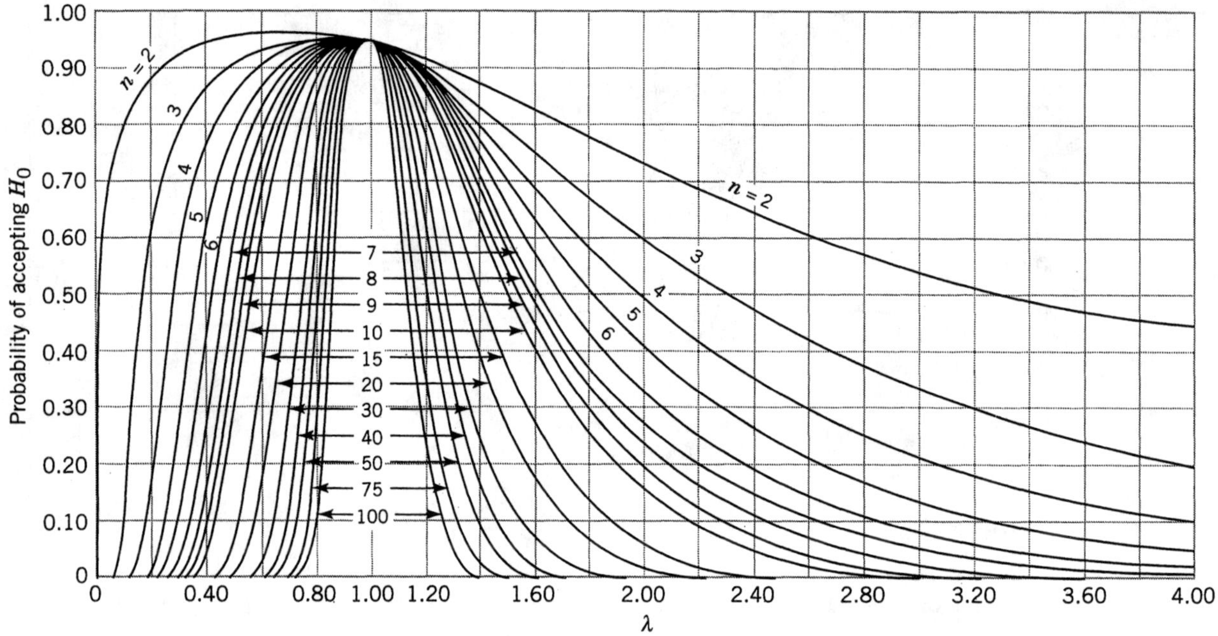

(*i*) *OC* curves for different values of *n* for the two-sided chi-square test for a level of significance $\alpha = 0.05$.

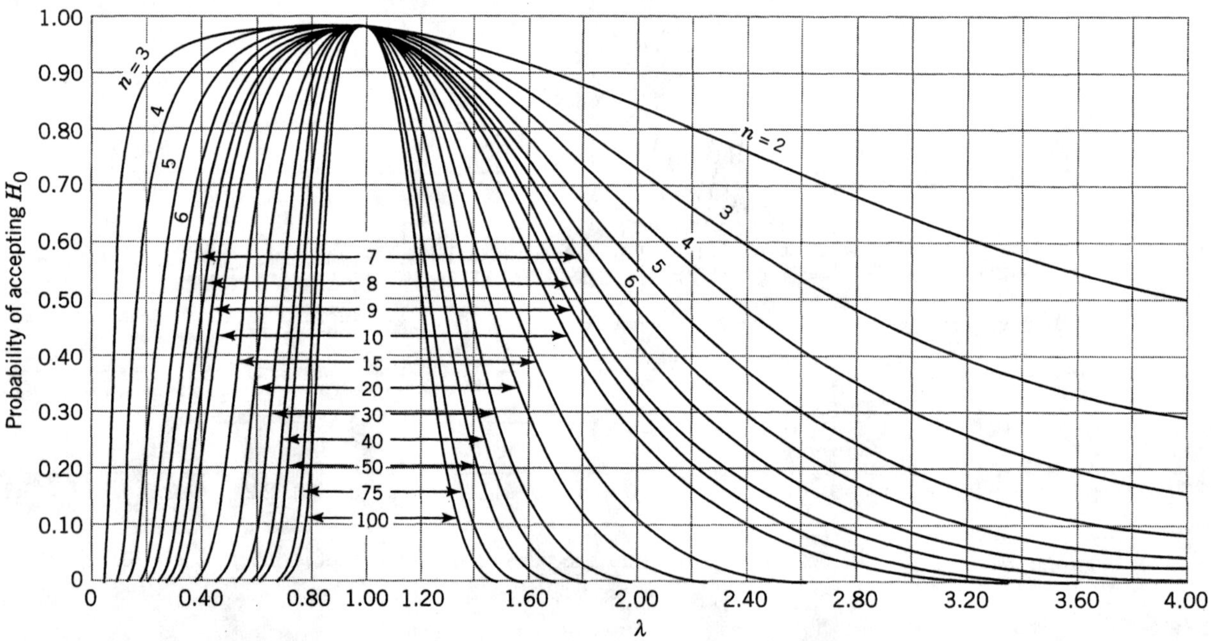

(*j*) *OC* curves for different values of *n* for the two-sided chi-square test for a level of significance $\alpha = 0.01$.

Chart VI Operating Characteristic Curves (*continued*)

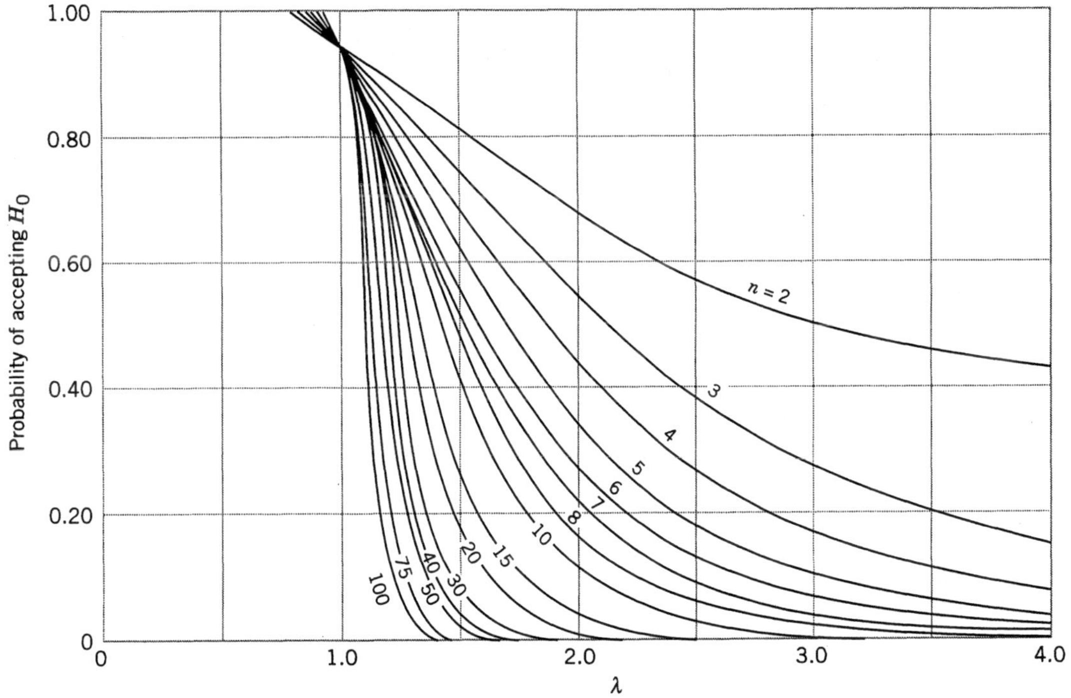

(*k*) *OC* curves for different values of *n* for the one-sided (upper tail) chi-square test for a level of significance $\alpha = 0.05$.

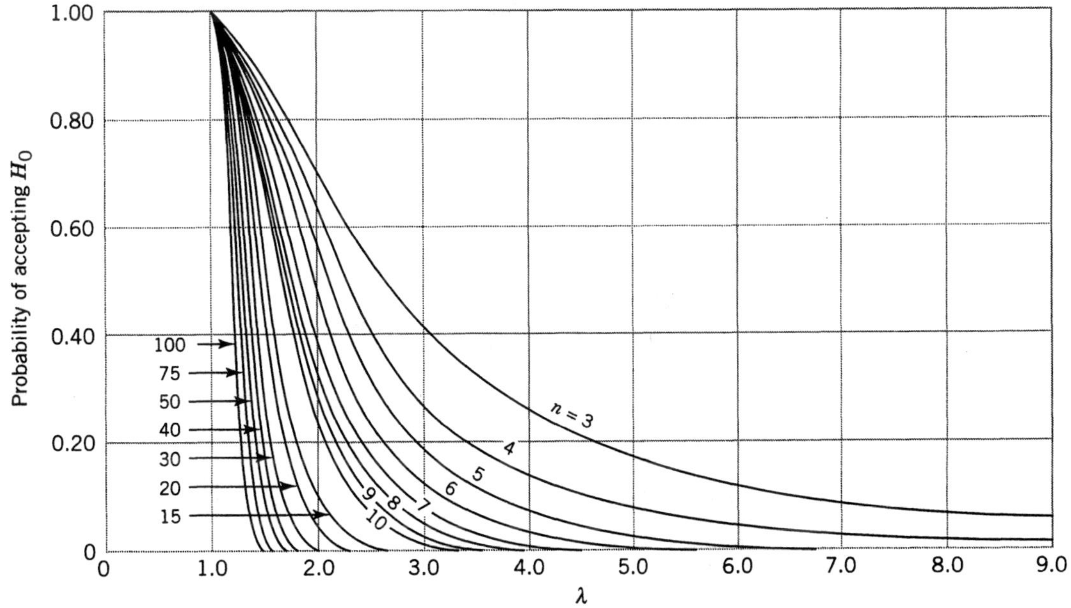

(*l*) *OC* curves for different values of *n* for the one-sided (upper tail) chi-square test for a level of significance $\alpha = 0.01$.

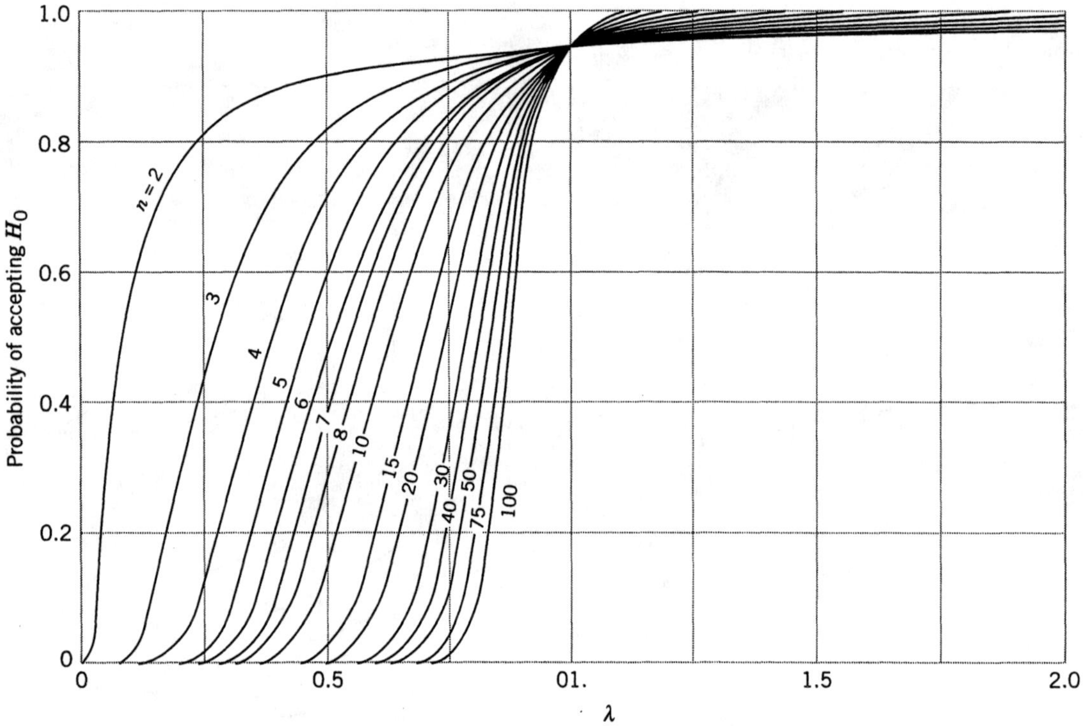

(*m*) *OC* curves for different values of *n* for the one-sided (lower tail) chi-square test for a level of significance $\alpha = 0.05$.

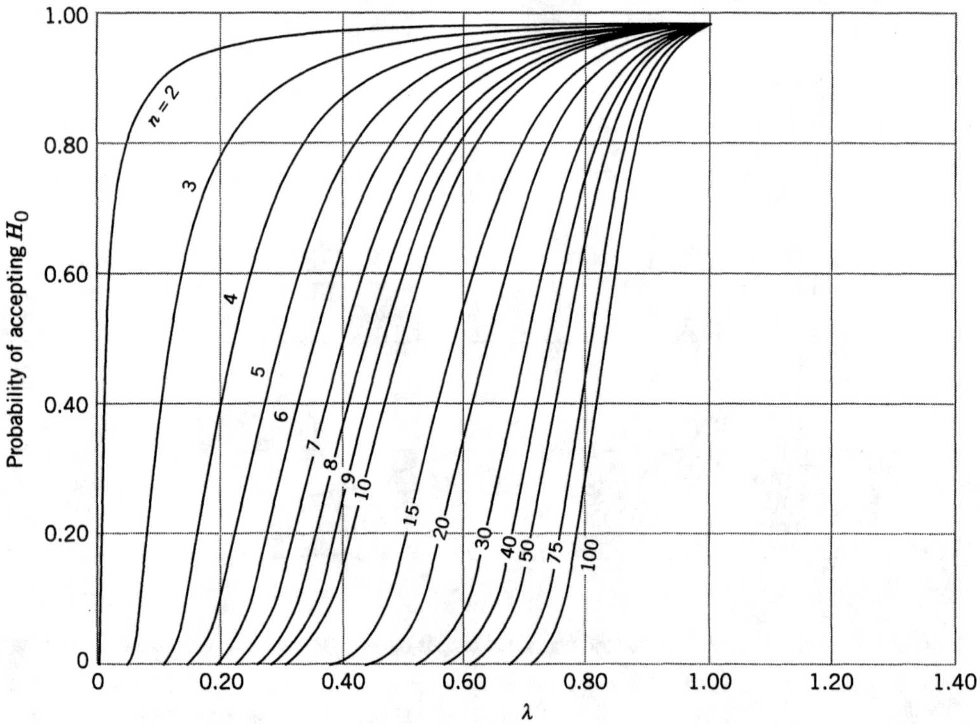

(*n*) *OC* curves for different values of *n* for the one-sided (lower tail) chi-square test for a level of significance $\alpha = 0.01$.

Chart VI Operating Characteristic Curves (*continued*)

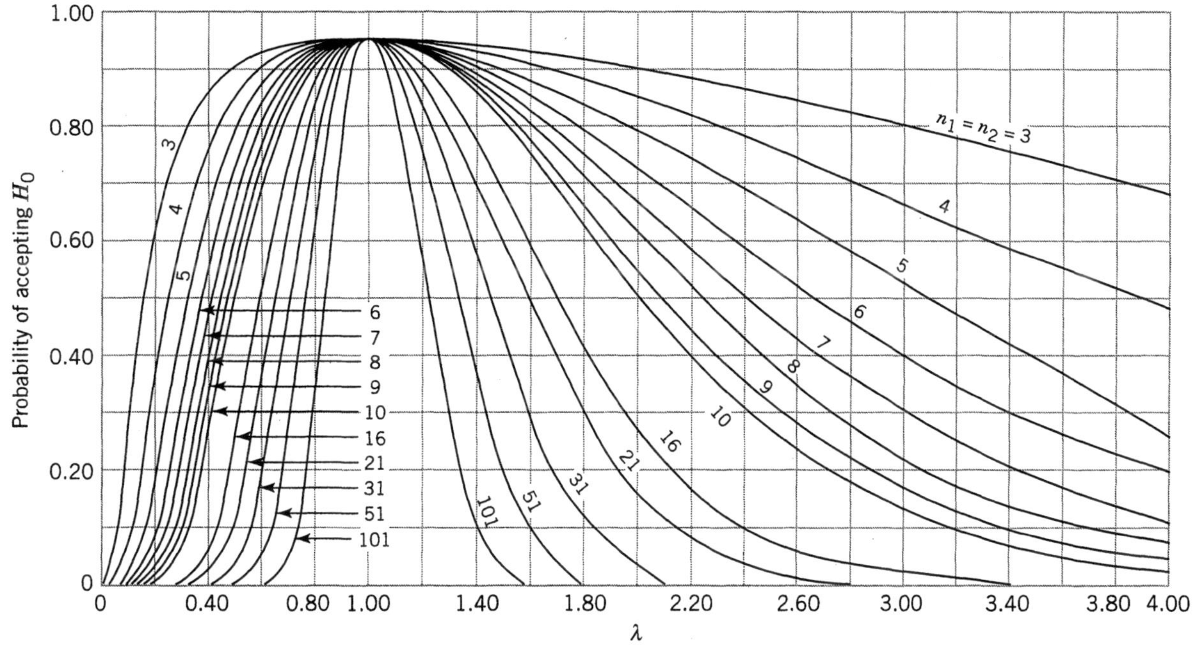

(*o*) *OC* curves for different values of *n* for the two-sided *F*-test for a level of significance $\alpha = 0.05$.

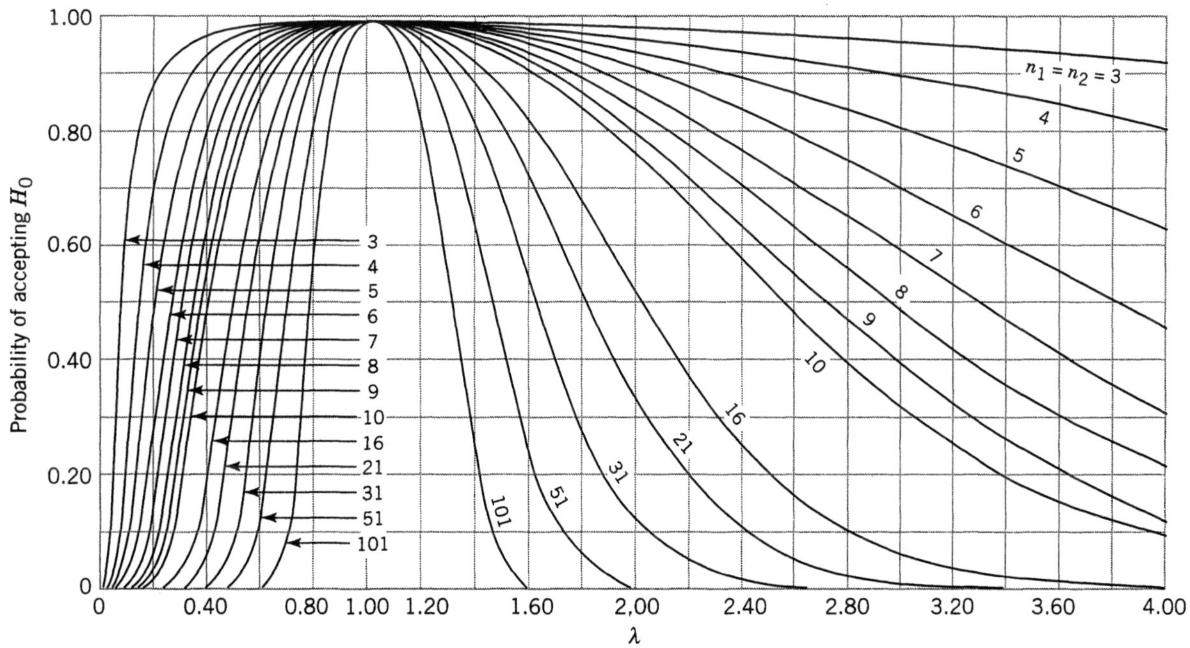

(*p*) *OC* curves for different values of *n* for the two-sided *F*-test for a level of significance $\alpha = 0.01$.

Chart VI Operating Characteristic Curves (*continued*)

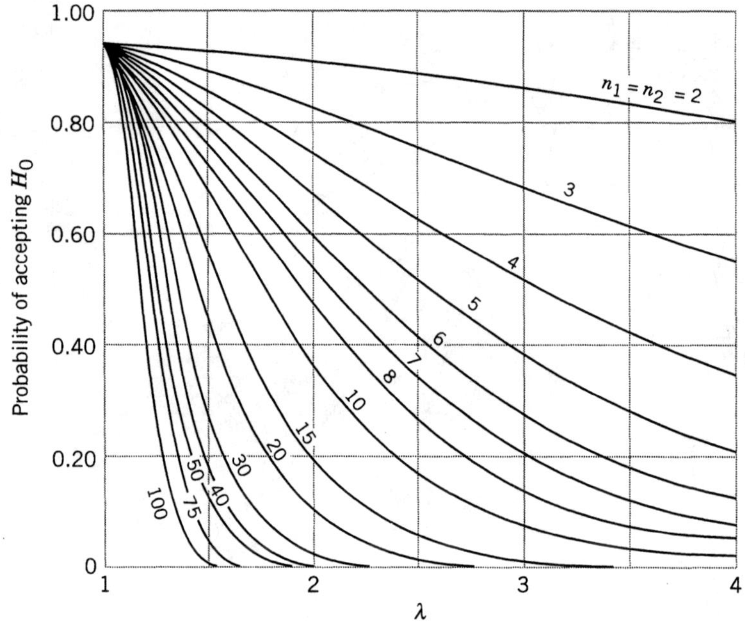

(*q*) *OC* curves for different values of *n* for the one-sided *F*-test for a level of significance $\alpha = 0.05$.

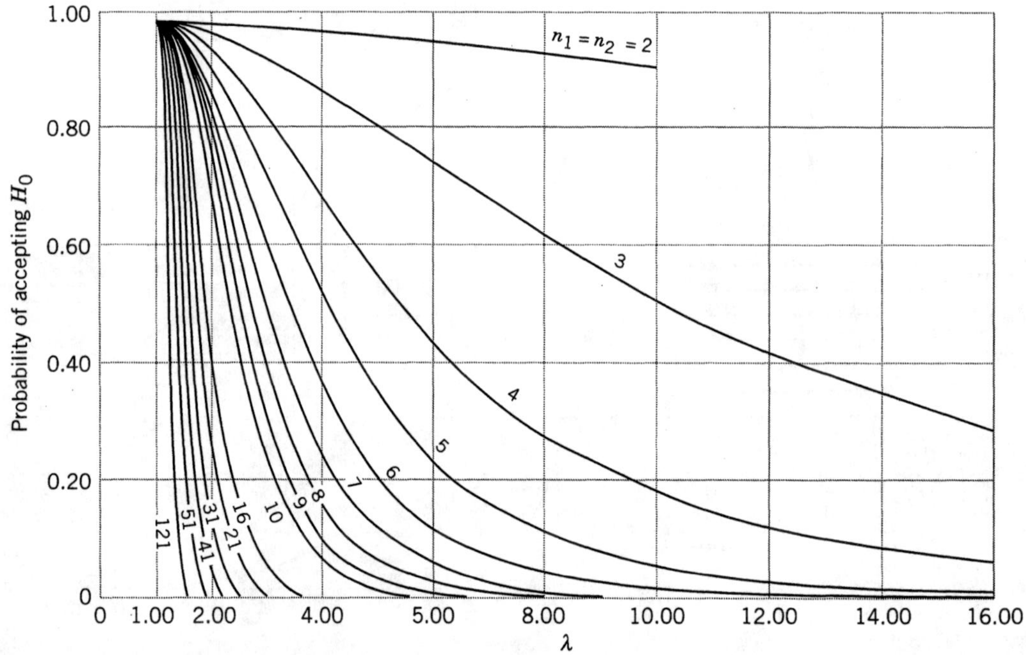

(*r*) *OC* curves for different values of *n* for the one-sided *F*-test for a level of significance $\alpha = 0.01$.

Chart VII Operating Characteristic Curves for the Fixed-Effects Model Analysis of Variance

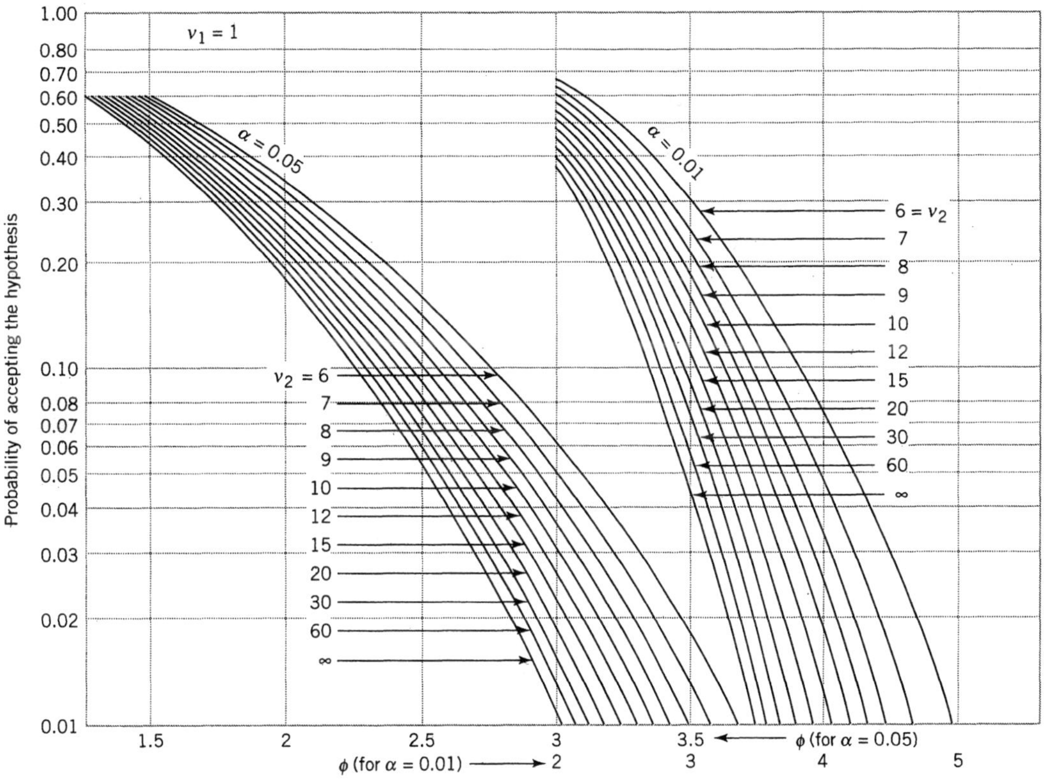

ν_1 = numerator degrees of freedom. ν_2 = denominator degrees of freedom.

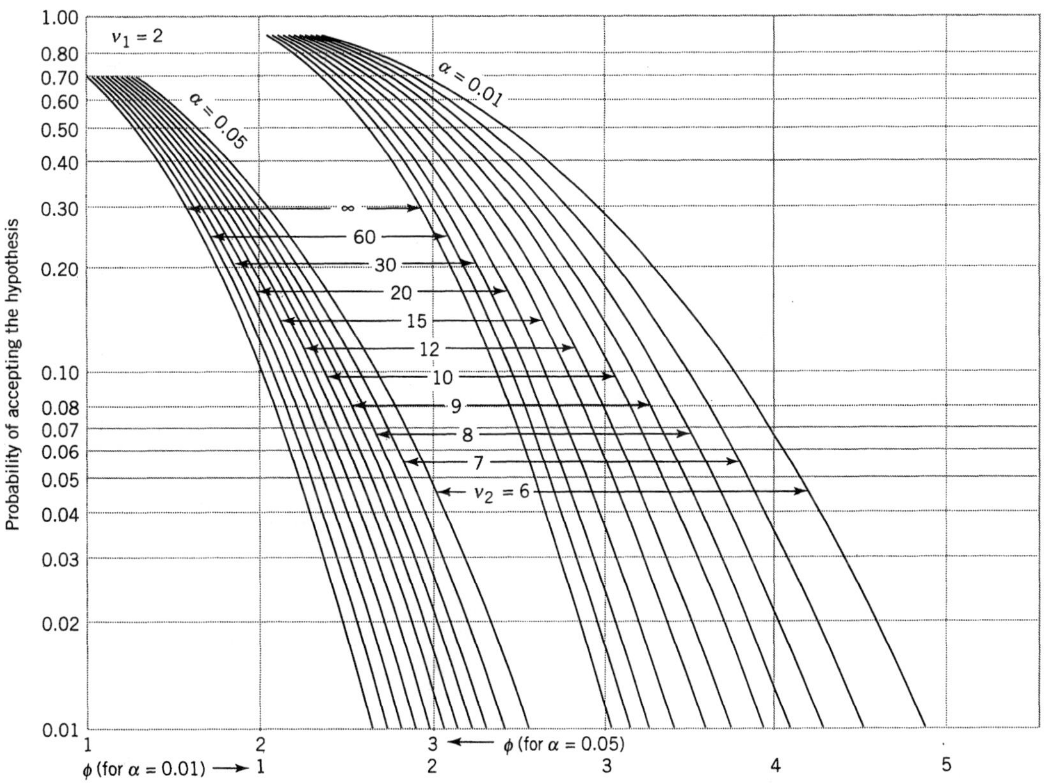

Source: Chart VII is adapted with permission from *Biometrika Tables for Statisticians,* Vol. 2, by E. S. Pearson and H. O. Hartley, Cambridge University Press, Cambridge, 1972.

Chart VII Operating Characteristic Curves for the Fixed-Effects Model Analysis of Variance (*continued*)

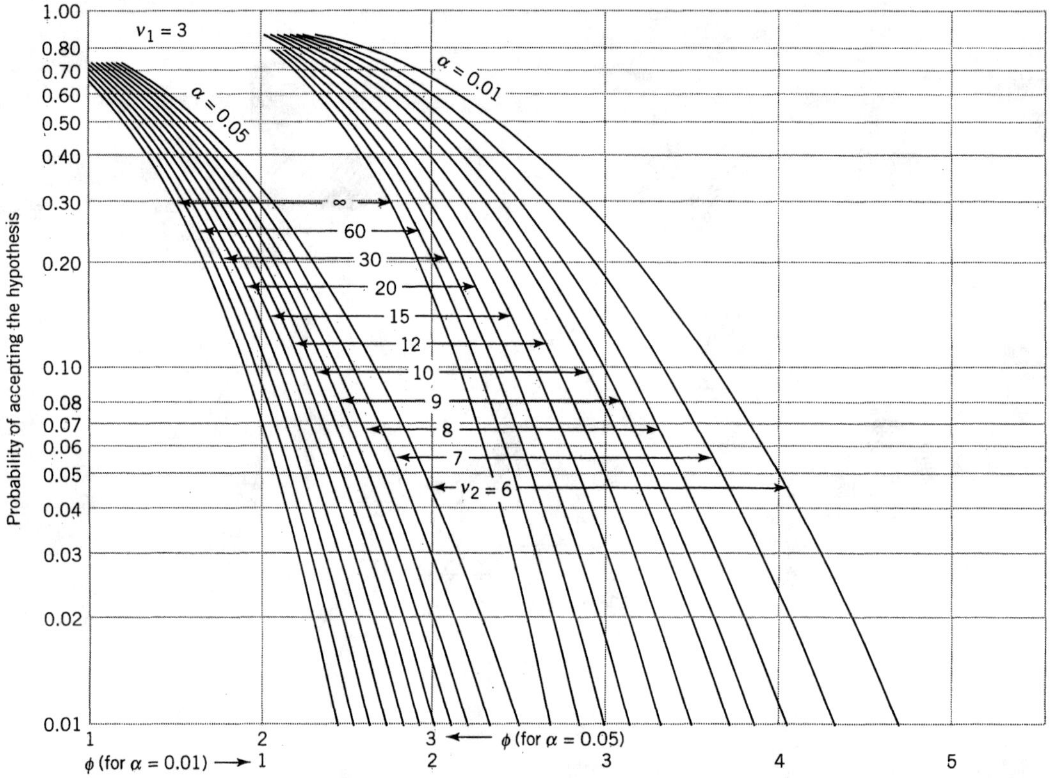

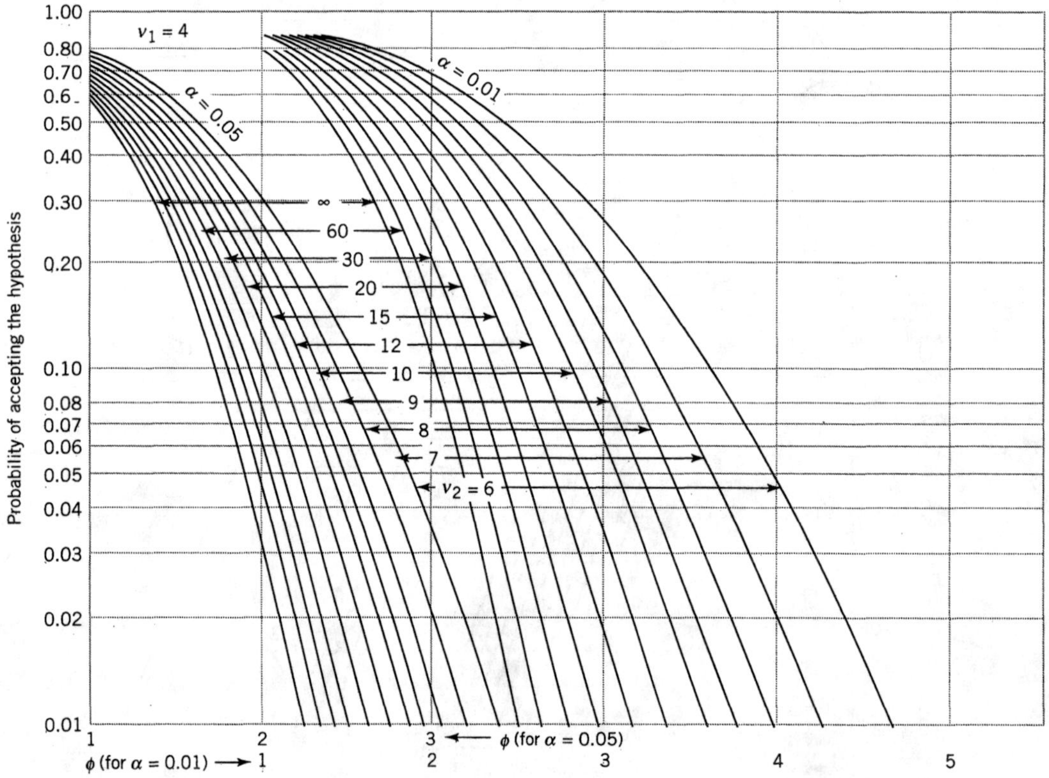

Chart VII Operating Characteristic Curves for the Fixed-Effects Model Analysis of Variance (*continued*)

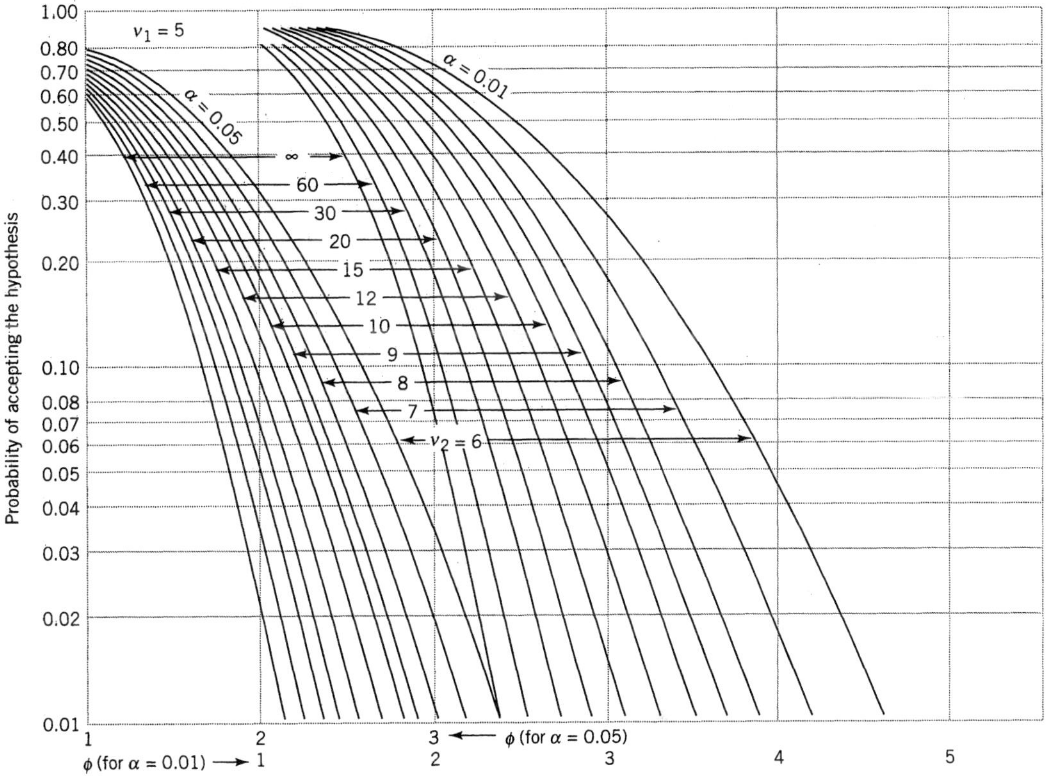

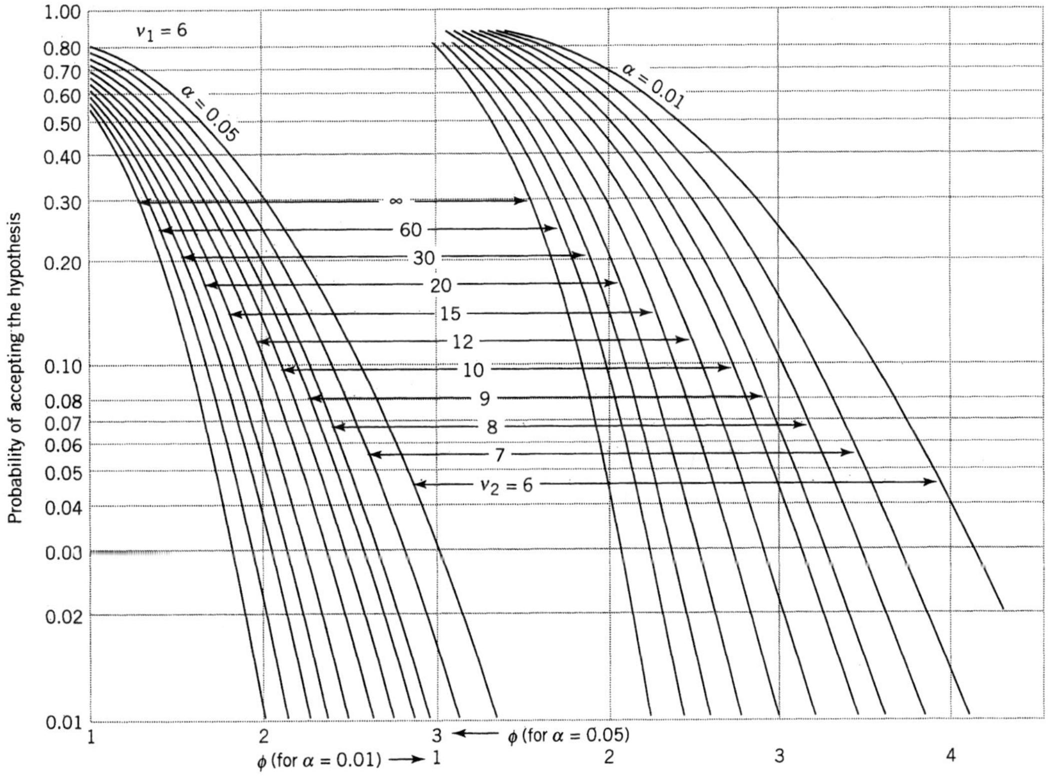

Chart VII Operating Characteristic Curves for the Fixed-Effects Model Analysis of Variance (*continued*)

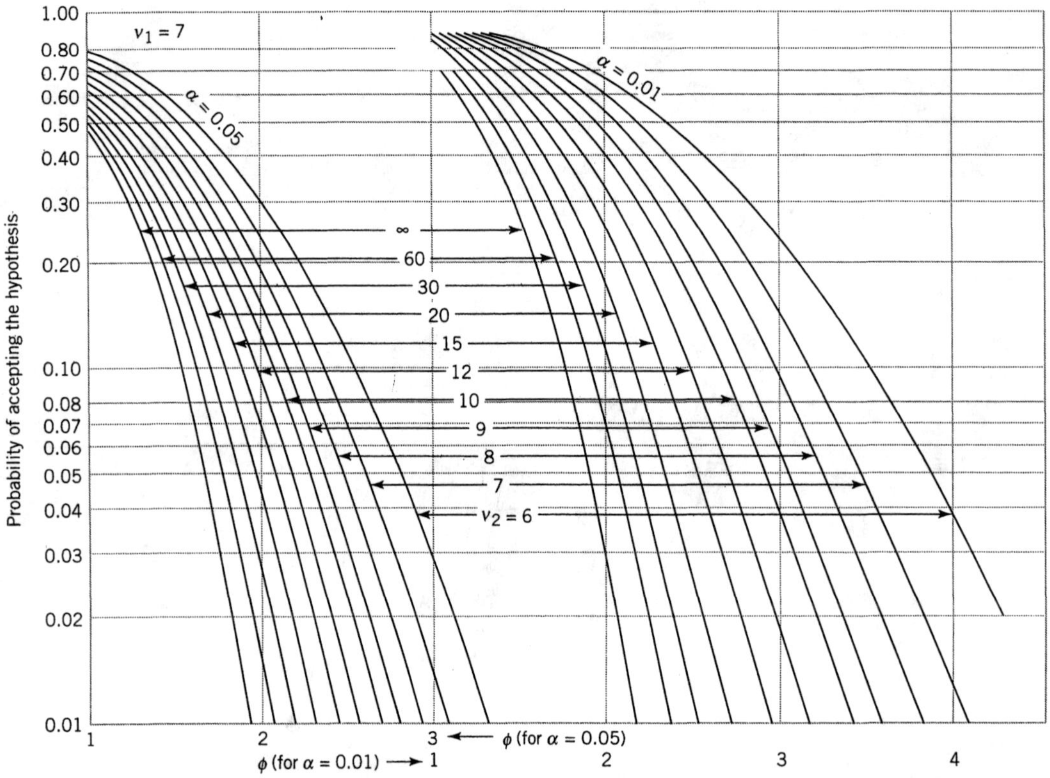

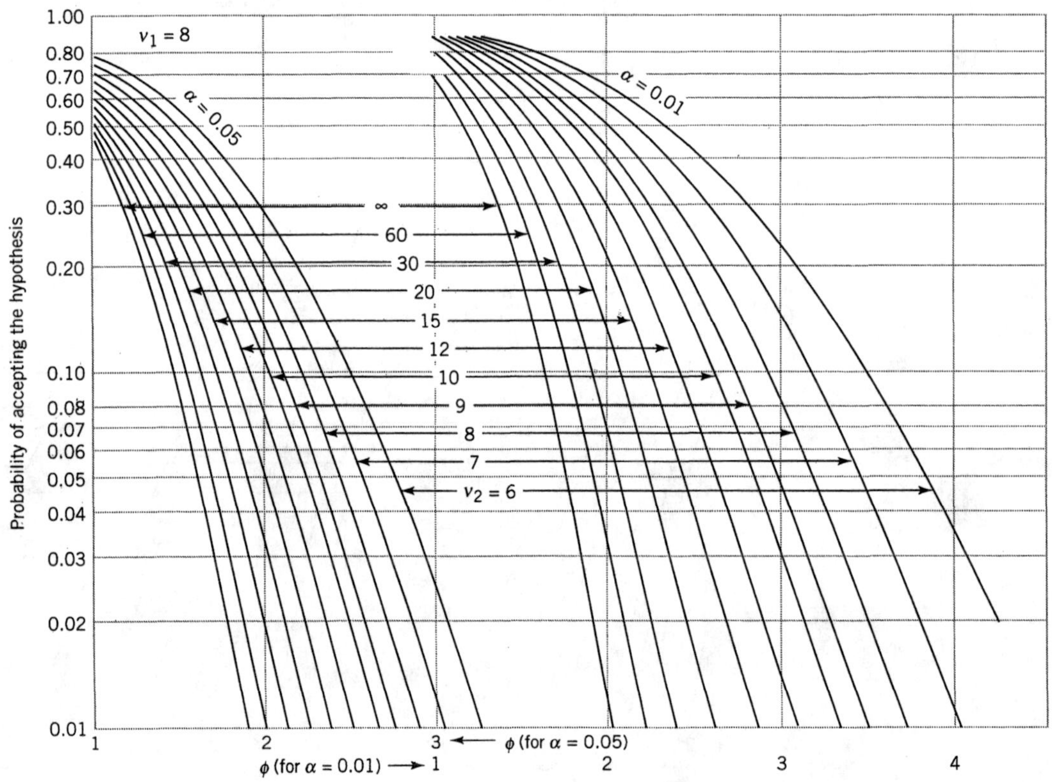

Chart VIII Operating Characteristic Curves for the Random-Effects Model Analysis of Variance

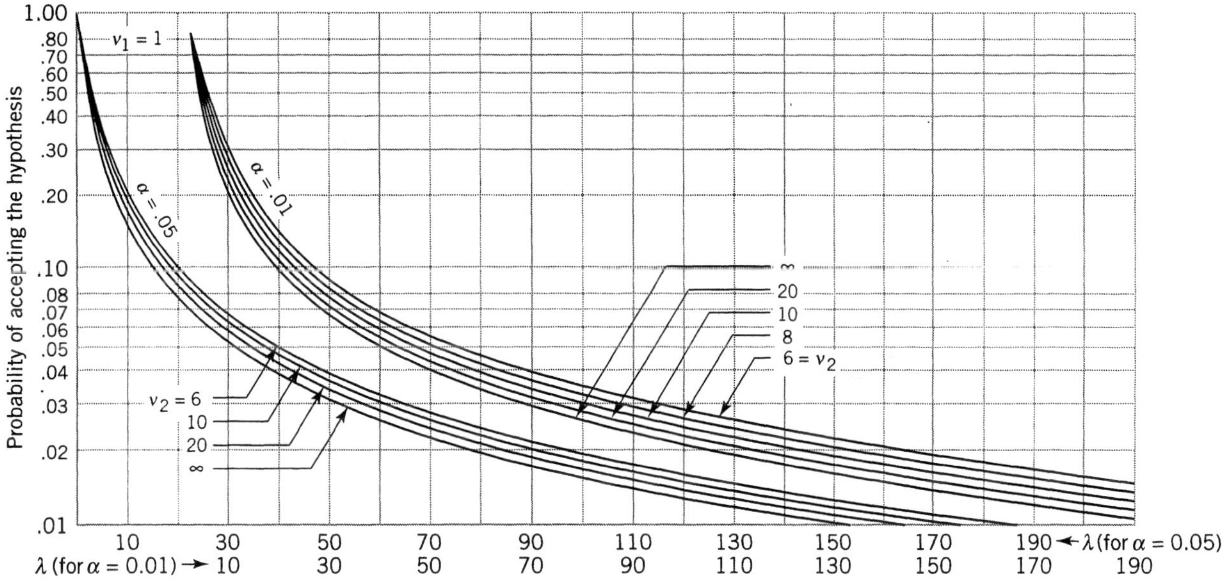

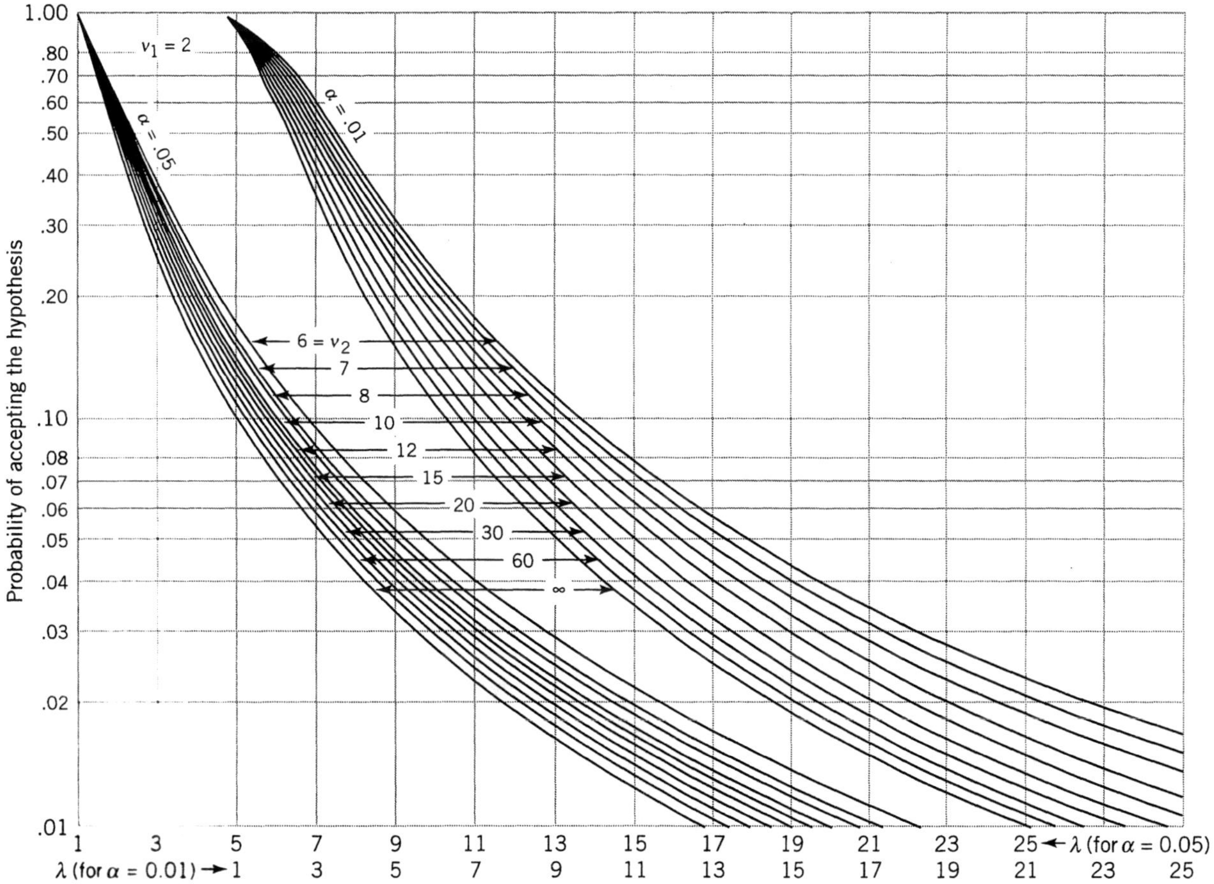

Source: Reproduced with permission from *Engineering Statistics,* 2nd edition, by A. H. Bowker and G. J. Lieberman. Prentice-Hall, Englewood Cliffs, N.J., 1972.

Chart VIII Operating Characteristic Curves for the Random-Effects Model Analysis of Variance (*continued*)

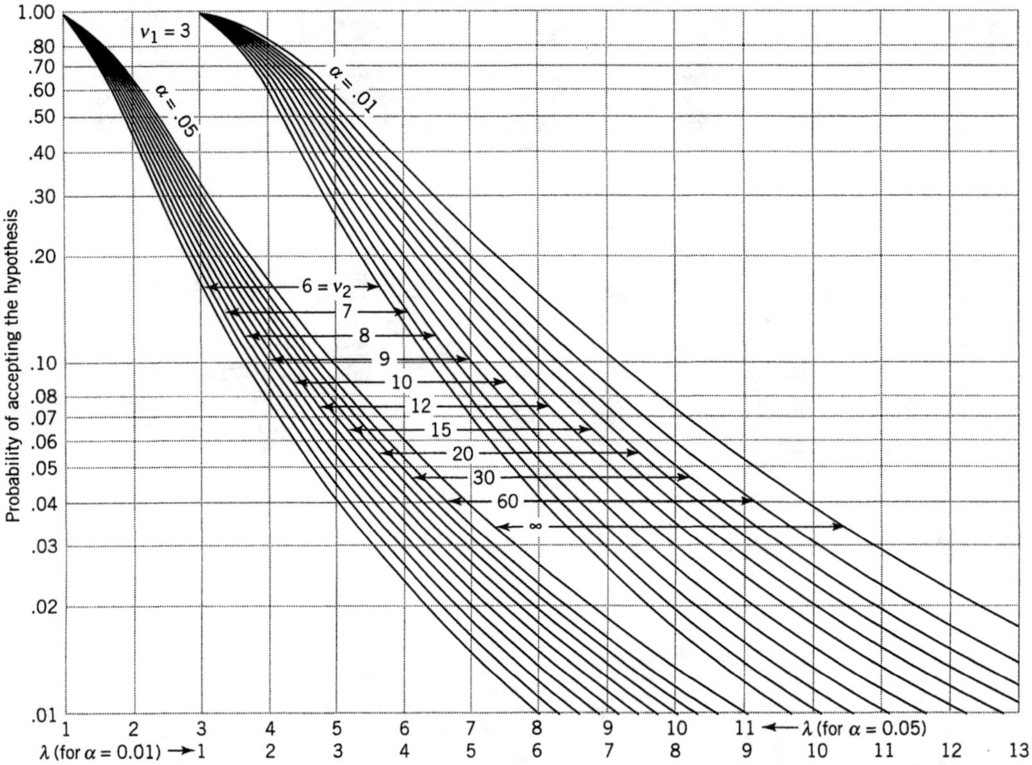

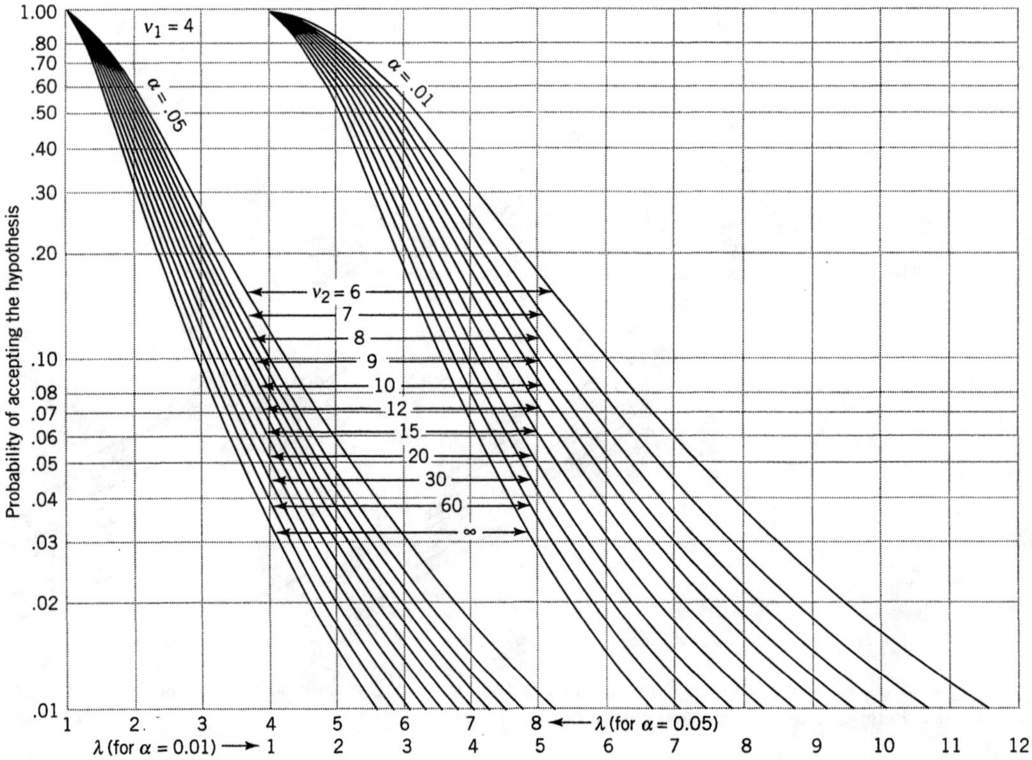

Chart VIII Operating Characteristic Curves for the Random-Effects Model Analysis of Variance (*continued*)

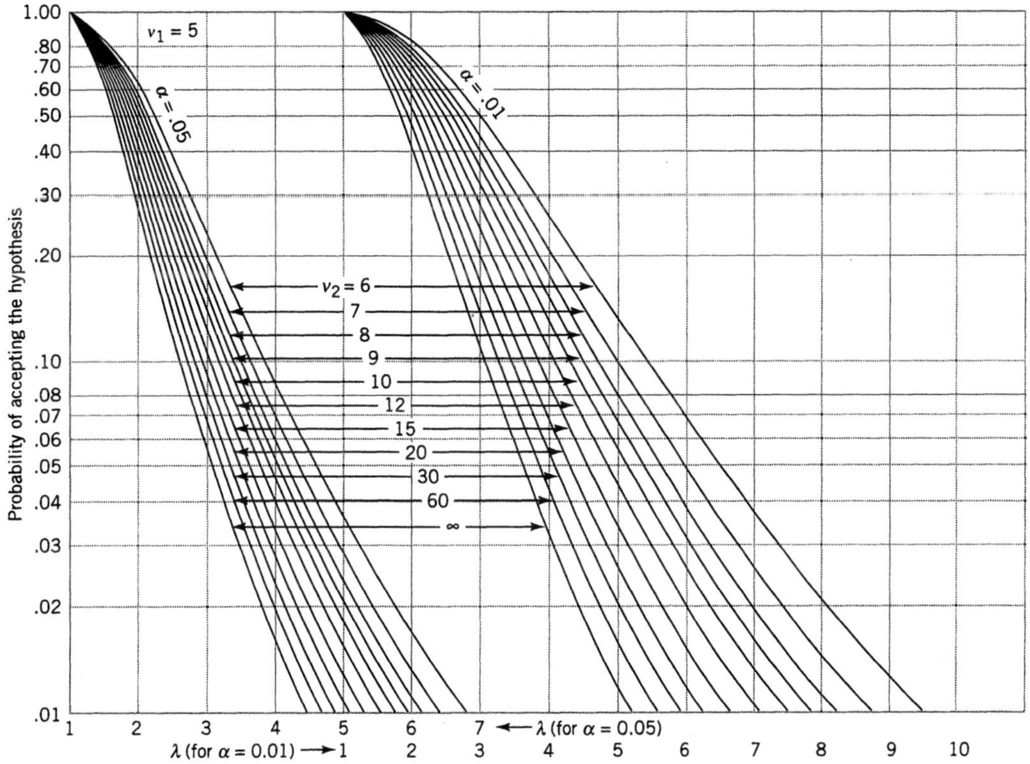

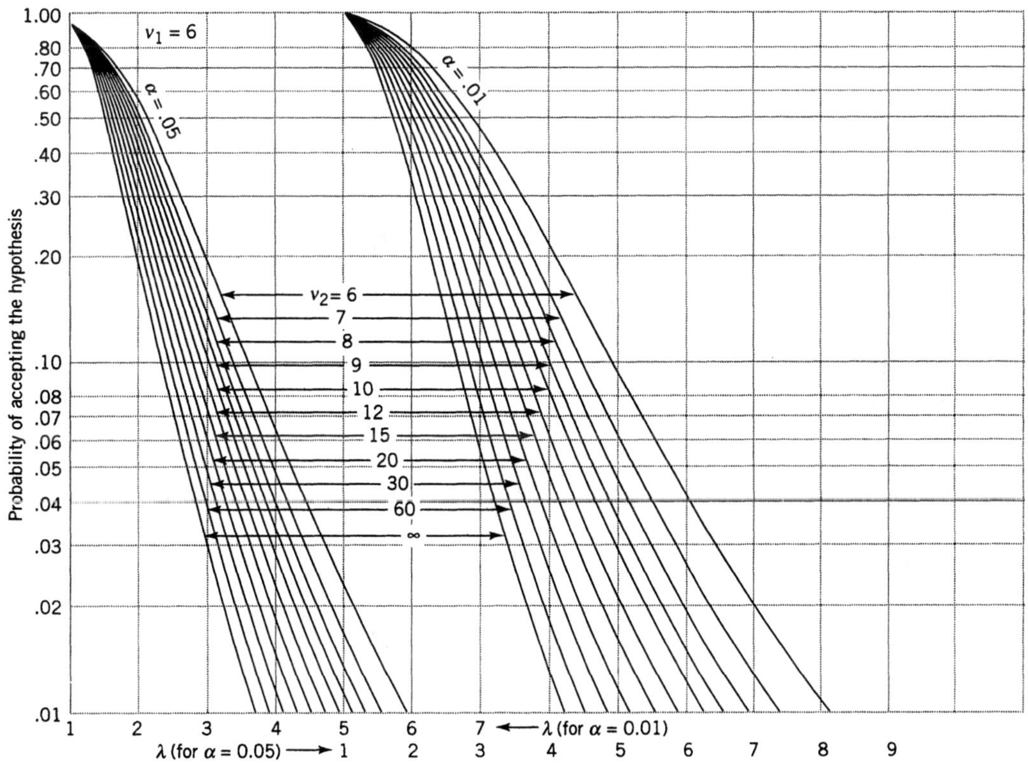

Chart VIII Operating Characteristic Curves for the Random-Effects Model Analysis of Variance (*continued*)

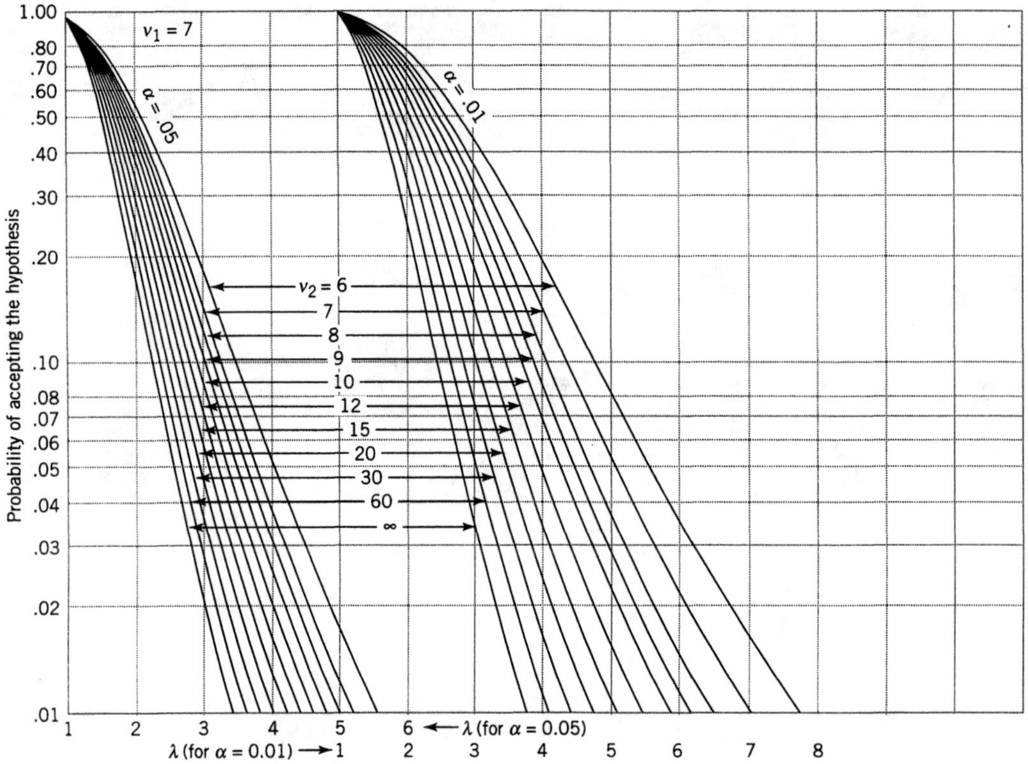

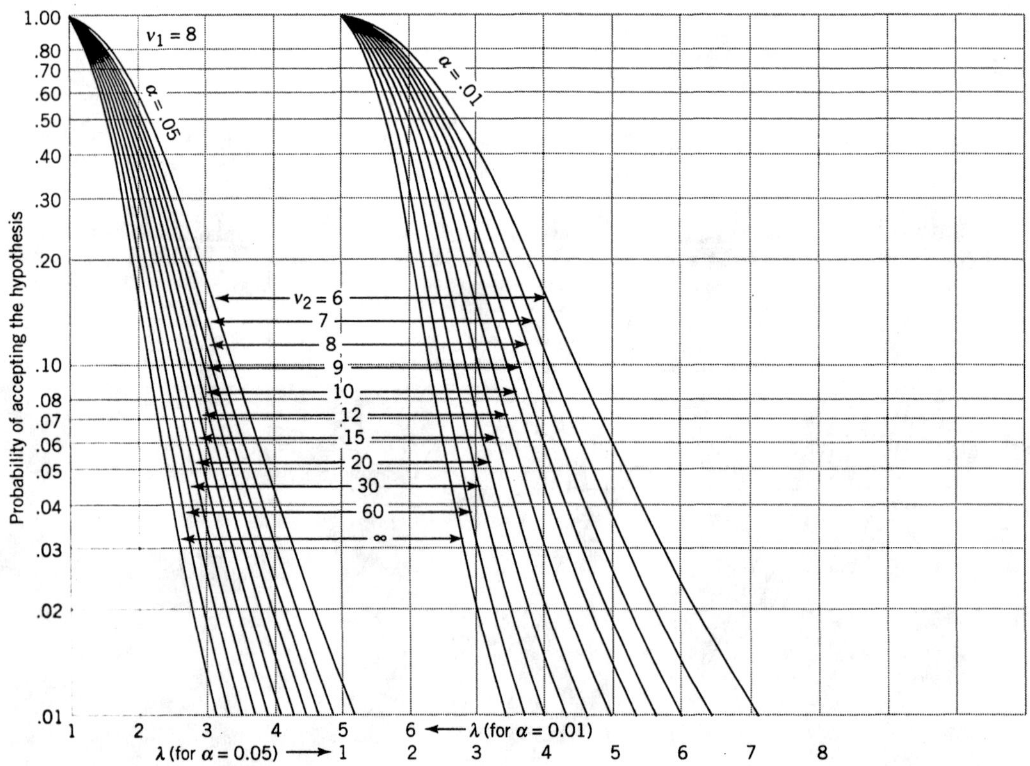

Table IX Critical Values for the Sign Test

r^*_α

n \ α	0.10	0.05	0.01	two-sided tests
	0.05	0.025	0.005	one-sided tests
5	0			
6	0	0		
7	0	0		
8	1	0	0	
9	1	1	0	
10	1	1	0	
11	2	1	0	
12	2	2	1	
13	3	2	1	
14	3	2	1	
15	3	3	2	
16	4	3	2	
17	4	4	2	
18	5	4	3	
19	5	4	3	
20	5	5	3	
21	6	5	4	
22	6	5	4	
23	7	6	4	
24	7	6	5	
25	7	7	5	
26	8	7	6	
27	8	7	6	
28	9	8	6	
29	9	8	7	
30	10	9	7	
31	10	9	7	
32	10	9	8	
33	11	10	8	
34	11	10	9	
35	12	11	9	
36	12	11	9	
37	13	12	10	
38	13	12	10	
39	13	12	11	
40	14	13	11	

Table X Critical Values for the Wilcoxon Signed-Rank Test

w^*_α

n^a \ α	0.10 0.05	0.05 0.025	0.02 0.01	0.01 0.005	two-sided tests one-sided tests
4					
5	0				
6	2	0			
7	3	2	0		
8	5	3	1	0	
9	8	5	3	1	
10	10	8	5	3	
11	13	10	7	5	
12	17	13	9	7	
13	21	17	12	9	
14	25	21	15	12	
15	30	25	19	15	
16	35	29	23	19	
17	41	34	27	23	
18	47	40	32	27	
19	53	46	37	32	
20	60	52	43	37	
21	67	58	49	42	
22	75	65	55	48	
23	83	73	62	54	
24	91	81	69	61	
25	100	89	76	68	

[a] If $n > 25$, W^+ (or W^-) is approximately normally distributed with mean $n(n + 1)/4$ and variance $n(n + 1)(2n + 1)/24$.

Table XI Critical Values for the Wilcoxon Rank-Sum Test

$w_{0.05}$

n_2 \ n_1[a]	4	5	6	7	8	9	10	11	12	13	14	15
4	10											
5	11	17										
6	12	18	26									
7	13	20	27	36								
8	14	21	29	38	49							
9	15	22	31	40	51	63						
10	15	23	32	42	53	65	78					
11	16	24	34	44	55	68	81	96				
12	17	26	35	46	58	71	85	99	115			
13	18	27	37	48	60	73	88	103	119	137		
14	19	28	38	50	63	76	91	106	123	141	160	
15	20	29	40	52	65	79	94	110	127	145	164	185
16	21	31	42	54	67	82	97	114	131	150	169	
17	21	32	43	56	70	84	100	117	135	154		
18	22	33	45	58	72	87	103	121	139			
19	23	34	46	60	74	90	107	124				
20	24	35	48	62	77	93	110					
21	25	37	50	64	79	95						
22	26	38	51	66	82							
23	27	39	53	68								
24	28	40	55									
25	28	42										
26	29											
27												
28												

[a]For n_1 and $n_2 > 8$, W_1 is approximately normally distributed with mean $\frac{1}{2}n_1(n_1 + n_2 + 1)$ and variance $n_1 n_2(n_1 + n_2 + 1)/12$.

Table XI Critical Values for the Wilcoxon Rank-Sum Test (*continued*)

$w_{0.01}$

n_2 \ n_1	4	5	6	7	8	9	10	11	12	13	14	15
5		15										
6	10	16	23									
7	10	17	24	32								
8	11	17	25	34	43							
9	11	18	26	35	45	56						
10	12	19	27	37	47	58	71					
11	12	20	28	38	49	61	74	87				
12	13	21	30	40	51	63	76	90	106			
13	14	22	31	41	53	65	79	93	109	125		
14	14	22	32	43	54	67	81	96	112	129	147	
15	15	23	33	44	56	70	84	99	115	133	151	171
16	15	24	34	46	58	72	86	102	119	137	155	
17	16	25	36	47	60	74	89	105	122	140		
18	16	26	37	49	62	76	92	108	125			
19	17	27	38	50	64	78	94	111				
20	18	28	39	52	66	81	97					
21	18	29	40	53	68	83						
22	19	29	42	55	70							
23	19	30	43	57								
24	20	31	44									
25	20	32										
26	21											
27												
28												

Table XII Factors for Constructing Variables Control Charts

	Factor for Control Limits					
	$\overline{X}$ Chart			R Chart		
n^a	A_1	A_2	d_2	D_3	D_4	n
2	3.760	1.880	1.128	0	3.267	2
3	2.394	1.023	1.693	0	2.575	3
4	1.880	.729	2.059	0	2.282	4
5	1.596	.577	2.326	0	2.115	5
6	1.410	.483	2.534	0	2.004	6
7	1.277	.419	2.704	.076	1.924	7
8	1.175	.373	2.847	.136	1.864	8
9	1.094	.337	2.970	.184	1.816	9
10	1.028	.308	3.078	.223	1.777	10
11	.973	.285	3.173	.256	1.744	11
12	.925	.266	3.258	.284	1.716	12
13	.884	.249	3.336	.308	1.692	13
14	.848	.235	3.407	.329	1.671	14
15	.816	.223	3.472	.348	1.652	15
16	.788	.212	3.532	.364	1.636	16
17	.762	.203	3.588	.379	1.621	17
18	.738	.194	3.640	.392	1.608	18
19	.717	.187	3.689	.404	1.596	19
20	.697	.180	3.735	.414	1.586	20
21	.679	.173	3.778	.425	1.575	21
22	.662	.167	3.819	.434	1.566	22
23	.647	.162	3.858	.443	1.557	23
24	.632	.157	3.895	.452	1.548	24
25	.619	.153	3.931	.459	1.541	25

[a] $n > 25$: $A_1 = 3/\sqrt{n}$ where n = number of observations in sample.

Appendix B Supplemental Technical Material

I. COUNTING TECHNIQUES

In many of the examples in Chapter 3, it is easy to determine the number of outcomes in each event. In more complicated examples, determining the number of outcomes that comprise the sample space (or an event) becomes more difficult. To associate probabilities with events, it is important to know the number of outcomes both in an event and in the sample space. Some simple rules can be used to simplify the calculations.

In Example 3-7, an automobile manufacturer provides vehicles equipped with selected options. Each vehicle is ordered

with or without an automatic transmission

with or without air conditioning

with one of three choices of a stereo system

with one of four exterior colors

The tree diagram in Fig. 3-7 describes the sample space of all possible vehicle types. The size of the sample space equals the number of branches in the last level of the tree and this quantity equals $2 \times 2 \times 3 \times 4 = 48$. This leads to the following useful result.

Multiplication Rule

If an operation can be described as a sequence of k steps, and

if the number of ways of completing step 1 is n_1, and

if the number of ways of completing step 2 is n_2 for each way of completing step 1, and

if the number of ways of completing step 3 is n_3 for each way of completing step 2, and so forth,

then the total number of ways of completing the operation is

$$n_1 \times n_2 \times \cdots \times n_k$$

EXAMPLE I-1

In the design of a casing for a gear housing, we can use four different types of fasteners, three different bolt lengths, and three different bolt locations. From the multiplication rule, $4 \times 3 \times 3 = 36$ different designs are possible.

Another useful calculation is the number of arrangements (orderings) of the elements of a set. Consider a set of elements, such as $S = \{a, b, c\}$. A **permutation** of the elements is an arrangement or an ordering of the elements. For example, *abc, acb, bac, bca, cab,* and *cba* are all of the permutations of the elements of *S*.

The number of **permutations** (arrangements) of n different elements is denoted as $n!$ and

$$n! = n \times (n-1) \times (n-2) \times \cdots \times 2 \times 1 \qquad \text{(I-1)}$$

This result follows from the multiplication rule. A permutation can be constructed by selecting the element to be placed in the first position of the arrangement from the n elements, then selecting the element for the second position from the $n - 1$ remaining elements, then selecting the element for the third position from the remaining $n - 2$ elements, and so forth. Permutations such as these are sometimes referred to as linear permutations.

In some situations, we are interested in the number of arrangements of only some of the elements of a set. The following result also follows from the multiplication rule.

The number of permutations (arrangements) of a subset of r elements selected from a set of n different elements is

$$P_r^n = n \times (n-1) \times (n-2) \times \cdots \times (n-r+1) = \frac{n!}{(n-r)!} \quad \text{(I-2)}$$

EXAMPLE I-2

A printed circuit board has eight different locations in which a component can be placed. If four different components are to be placed on the board, how many different designs are possible?

Each design consists of selecting a location from the eight locations for the first component, a location from the remaining seven for the second component, a location from the remaining six for the third component, and a location from the remaining five for the fourth component. Therefore,

$$P_4^8 = 8 \times 7 \times 6 \times 5 = \frac{8!}{4!} = 1680 \text{ different designs are possible.}$$

Sometimes we are interested in counting the number of arrangements or ordering of objects that are not all different. The following result is a useful, general calculation.

The number of arrangements of $n = n_1 + n_2 + \cdots + n_r$ objects of which

n_1 are of one type,

n_2 are of a second type,

n_3 are of a third type,

. . . , and

n_r are of an rth type is

$$\frac{n!}{n_1!\, n_2!\, n_3! \ldots n_r!} \quad \text{(I-3)}$$

EXAMPLE I-3

Consider a machining operation in which a piece of sheet metal needs two different diameter holes drilled and two different size notches cut. We denote a drilling operation as d and a notching operation as n. In determining a schedule for a machine shop, we might be interested in the number of different possible sequences of the four operations. The number of possible sequences for two drilling operations and two notching operations is

$$\frac{4!}{2!\,2!} = 6$$

The six sequences are easily summarized: *ddnn, dndn, dnnd, nddn, ndnd, nndd.*

EXAMPLE I-4

A part is labeled by printing with four thick lines, three medium lines, and two thin lines. If each ordering of the nine lines represents a different label, how many different labels can be generated by using this scheme?

From equation I-3, the number of possible part labels is

$$\frac{9!}{4!\,3!\,2!} = 2520$$

Another counting problem of interest is the number of subsets of r elements that can be selected from a set of n elements. Here, order is not important. Every subset of r elements can be indicated by listing the elements in the set and marking each element with a "*" if it is to be included in the subset. Therefore, each arrangement of r "*"'s and $n - r$ blanks indicates a different subset. From equation I-3, we obtain the following result.

> The number of subsets of size r that can be selected from a set of n elements is denoted as $\binom{n}{r}$ or C_r^n and
>
> $$\binom{n}{r} = \frac{n!}{r!(n-r)!} \tag{I-4}$$

EXAMPLE I-5

A printed circuit board has eight different locations in which a component can be placed. If five identical components are to be placed on the board, how many different designs are possible?

Each design is a subset of the eight locations that are to contain the components. From equation I-4, the number of possible designs is

$$\frac{8!}{5!\,3!} = 56$$

The following example uses the multiplication rule in combination with equation I-4 to answer a more difficult, but common, question.

EXAMPLE I-6

A bin of 50 manufactured parts contains three defective parts and 47 non-defective parts. A sample of six parts is selected from the 50 parts. Selected parts are not replaced. That is, each part can only be selected once and the sample is a subset of the 50 parts. How many different samples are there of size six that contain exactly two defective parts?

A subset containing exactly two defective parts can be formed by first choosing the two defective parts from the three defective parts. Using equation I-4, this step can be completed in

$$\frac{3!}{2!\,1!} = 3 \text{ different ways}$$

Then, the second step is to select the remaining four parts from the 47 acceptable parts in the bin. The second step can be completed in

$$\frac{47!}{4!\,43!} = 178{,}365 \text{ different ways}$$

Therefore, from the multiplication rule, the number of subsets of size six that contain exactly two defective items is

$$3 \times 178{,}365 = 535{,}095$$

As an additional computation, the total number of different subsets of size six is found to be

$$\frac{50!}{6!\,44!} = 15{,}890{,}700$$

The parts in Example I-6 are chosen without replacing previously selected parts. Consequently, each part can only be selected once and the sample is a subset of the set of all parts. This method of selection is often referred to as sampling **without replacement.** In Chapter 3 we stated that the selection of a part is said to be random if each part is equally likely. When more than one part is selected, the sampling is said to be **random** if each subset of parts is equally likely to be selected.

EXAMPLE I-7

In Example I-6, if the sampling is random, then each subset of six parts is equally likely. Therefore, the probability that a sample contains exactly two defective parts is

$$\frac{535{,}095}{15{,}890{,}700} = 0.034$$

Note that Example I-7 illustrates the hypergeometric distribution.

Exercises for Appendix I

I-1. An order for a personal digital assistant can specify any one of five memory sizes, any one of three types of displays, any one of four sizes of a hard disk, and can either include or not include a pen tablet. How many different systems can be ordered?

I-2. In a manufacturing operation, a part is produced by machining, polishing, and painting. If there are three machine tools, four polishing tools, and three painting tools, how many different routings (consisting of machining, followed by polishing, and followed by painting) for a part are possible?

I-3. New designs for a wastewater treatment tank have proposed three possible shapes, four possible sizes, three locations for input valves, and four locations for output valves. How many different product designs are possible?

I-4. A manufacturing process consists of 10 operations that can be completed in any order. How many different production sequences are possible?

I-5. A manufacturing operation consists of 10 operations. However, five machining operations must be completed before any of the remaining five assembly operations can begin. Within each set of five, operations can be completed in any order. How many different production sequences are possible?

I-6. In a sheet metal operation, three notches and four bends are required. If the operations can be done in any order, how many different ways of completing the manufacturing are possible?

I-7. A lot of 140 semiconductor chips is inspected by choosing a sample of five chips. Assume 10 of the chips do not conform to customer requirements.

(a) How many different samples are possible?

(b) How many samples of five contain exactly one nonconforming chip?

(c) How many samples of five contain at least one nonconforming chip?

I-8. In the layout of a printed circuit board for an electronic product, there are 12 different locations that can accommodate chips.

(a) If five different types of chips are to be placed on the board, how many different layouts are possible?

(b) If the five chips that are placed on the board are of the same type, how many different layouts are possible?

I-9. In the laboratory analysis of samples from a chemical process, five samples from the process are analyzed daily. In addition, a control sample is analyzed two times each day to check the calibration of the laboratory instruments.

(a) How many different sequences of process and control samples are possible each day? Assume that the five process samples are considered identical and that the two control samples are considered identical.

(b) How many different sequences of process and control samples are possible if we consider the five process samples to be different and the two control samples to be identical.

(c) For the same situation as part (b), how many sequences are possible if the first test of each day must be a control sample?

I-10. In the design of an electromechanical product, seven different components are to be stacked into a cylindrical casing that holds 12 components in a manner that minimizes the impact of shocks. One end of the casing is designated as the bottom and the other end is the top.

(a) How many different designs are possible?

(b) If the seven components are all identical, how many different designs are possible?

(c) If the seven components consist of

three of one type of component and four of another type, how many different designs are possible? (more difficult)

I-11. The design of a communication system considered the following questions.

(a) How many three-digit phone prefixes that are used to represent a particular geographic area (such as an area code) can be created from the digits 0 through 9?

(b) As in part (a), how many three-digit phone prefixes are possible that do not start with 0 or 1, but contain 0 or 1 as the middle digit?

(c) How many three-digit phone prefixes are possible in which no digit appears more than once in each prefix?

I-12. A byte is a sequence of eight bits and each bit is either 0 or 1.

(a) How many different bytes are possible?

(b) If the first bit of a byte is a parity check, that is, the first byte is determined from the other seven bits, how many different bytes are possible?

I-13. In a chemical plant, 24 holding tanks are used for final product storage. Four tanks are selected at random and without replacement. Suppose that six of the tanks contain material in which the viscosity exceeds the customer requirements.

(a) What is the probability that exactly one tank in the sample contains high viscosity material?

(b) What is the probability that at least one tank in the sample contains high viscosity material?

(c) In addition to the six tanks with high viscosity levels, four different tanks contain material with high impurities. What is the probability that exactly one tank in the sample contains high viscosity material and exactly one tank in the sample contains material with high impurities?

I-14. Plastic parts produced by an injection-molding operation are checked for conformance to specifications. Each tool contains 12 cavities in which parts are produced, and these parts fall into a conveyor when the press opens. An inspector chooses three parts from among the 12 at random. Two cavities are affected by a temperature malfunction that results in parts that do not conform to specifications.

(a) What is the probability that the inspector finds exactly one nonconforming part?

(b) What is the probability that the inspector finds at least one nonconforming part?

I-15. A bin of 50 parts contains five that are defective. A sample of two is selected at random, without replacement.

(a) Determine the probability that both parts in the sample are defective by computing a conditional probability as in Chapter 2.

(b) Determine the answer to part (a) by using the subset approach that was described in this section.

II. MOMENT GENERATING FUNCTION

Suppose that X is a random variable with mean μ. Throughout this book we have used the idea of the expected value of the random variable X, and in fact $E(X) = \mu$. Now suppose that we are interested in the expected value of a particular function of X, say, $g(X) = X^r$. The expected value of this function, or $E[g(X)] = E(X^r)$, is called the rth moment about the origin of the random variable X, which we will denote by μ_r'.

Definition

The rth **moment about the origin** of the random variable X is

$$\mu_r' = E(X^r) = \begin{cases} \sum_x x^r f(x), & X \text{ discrete} \\ \int_{-\infty}^{\infty} x^r f(x)\, dx, & X \text{ continuous} \end{cases} \tag{II-1}$$

Notice that the first moment about the origin is just the mean, that is, $E(X) = \mu_1' = \mu$. Furthermore, since the second moment about the origin is $E(X^2) = \mu_2'$, we can write the variance of a random variable in terms of origin moments as follows:

$$\begin{aligned} \sigma^2 &= E(X^2) - [E(X)]^2 \\ &= \mu_2' - \mu^2 \end{aligned}$$

The moments of a random variable can often be determined directly from the definition in equation II-1, but there is an alternative procedure that is frequently useful that makes use of a special function.

Definition

The **moment generating function** of the random variable X is the expected value of e^{tX} and is denoted by $M_X(t)$. That is,

$$M_X(t) = E(e^{tX}) = \begin{cases} \sum_x e^{tx} f(x), & X \text{ discrete} \\ \int_{-\infty}^{\infty} e^{tx} f(x)\, dx, & X \text{ continuous} \end{cases}$$

The moment generating function $M_X(t)$ will exist only if the sum or integral in the above definition converges. If the moment generating function of a random variable does exist, it can be used to obtain all the origin moments of the random variable.

Theorem II-1

Let X be a random variable with moment generating function $M_X(t)$. Then

$$\mu'_r = \left.\frac{d^r M_X(t)}{dt^r}\right|_{t=0}$$

Proof Assuming that we can differentiate inside the summation and integral signs, then

$$\frac{d^r M_X(t)}{dt^r} = \begin{cases} \sum_x x^r e^{tx} f(x), & X \text{ discrete} \\ \int_{-\infty}^{\infty} x^r e^{tx} f(x)\, dx, & X \text{ continuous} \end{cases} \tag{II-2}$$

Now if we set $t = 0$ in equation II-2, we find that

$$\left.\frac{d^r M_X(t)}{dt^r}\right|_{t=0} = E(X^r)$$

EXAMPLE II-1

Suppose that X has a binomial distribution, that is

$$\begin{aligned} f(x) &= \binom{n}{x} p^x (1-p)^{n-x}, & x &= 0, 1, \ldots, n \\ &= 0, & &\text{otherwise} \end{aligned}$$

Find the moment generating function and use it to verify that the mean and variance of the binomial random variable are $\mu = np$ and $\sigma^2 = np(1-p)$.

From the definition of a moment generating function, we have

$$\begin{aligned} M_X(t) &= \sum_{x=0}^{n} e^{tx} \binom{n}{x} p^x (1-p)^{n-x} \\ &= \sum_{x=0}^{n} \binom{n}{x} (pe^t)^x (1-p)^{n-x} \end{aligned}$$

This last summation is the binomial expansion of $[pe^t + (1-p)]^n$, so

$$M_X(t) = [pe^t + (1-p)]^n$$

Taking the first and second derivatives, we obtain

$$M'_X(t) = \frac{dM_X(t)}{dt} = npe^t[1 + p(e^t - 1)]^{n-1}$$

and

$$M''_X(t) = \frac{d^2M_X(t)}{dt^2} = npe^t(1 - p + npe^t)[1 + p(e^t - 1)]^{n-2}$$

If we set $t = 0$ in $M'_X(t)$, we obtain

$$M'_X(t)\big|_{t=0} = \mu'_1 = \mu = np$$

which is the mean of the binomial random variable X. Now if we set $t = 0$ in $M''_X(t)$, then

$$M''_X(t)\big|_{t=0} = \mu'_2 = np(1 - p + np)$$

Therefore, the variance of the binomial random variable is

$$\begin{aligned}\sigma^2 &= \mu'_2 - \mu^2 \\ &= np(1 - p + np) - (np)^2 \\ &= np - np^2 \\ &= np(1 - p)\end{aligned}$$

EXAMPLE II-2

Find the moment generating function of the normal random variable and use it to show that the mean and variance of this random variable are μ and σ^2, respectively.

The moment generating function is

$$\begin{aligned}M_X(t) &= \int_{-\infty}^{\infty} e^{tx} \frac{1}{\sigma\sqrt{2\pi}} e^{-(x-\mu)^2/(2\sigma^2)}\, dx \\ &= \int_{-\infty}^{\infty} \frac{1}{\sigma\sqrt{2\pi}} e^{-[x^2-2(\mu+t\sigma^2)x+\mu^2]/(2\sigma^2)}\, dx\end{aligned}$$

If we complete the square in the exponent, we have

$$x^2 - 2(\mu + t\sigma^2)x + \mu^2 = [x - (\mu + t\sigma^2)]^2 - 2\mu t\sigma^2 - t^2\sigma^4$$

and then

$$M_X(t) = \int_{-\infty}^{\infty} \frac{1}{\sigma\sqrt{2\pi}} e^{-\{[x-(\mu+t\sigma^2)]^2 - 2\mu t\sigma^2 - t^2\sigma^4\}/(2\sigma^2)}\, dx$$

$$= e^{\mu t + \sigma^2 t^2/2} \int_{-\infty}^{\infty} \frac{1}{\sigma\sqrt{2\pi}} e^{-(1/2)[x-(\mu+t\sigma^2)]^2/\sigma^2}\, dx$$

Let $u = [x - (\mu + t\sigma^2)]/\sigma$. Then $dx = \sigma\, du$, and this last expression above becomes

$$M_X(t) = e^{\mu t + \sigma^2 t^2/2} \int_{-\infty}^{\infty} \frac{1}{\sqrt{2\pi}} e^{-u^2/2}\, du$$

Now the integral is just the total area under a standard normal density, which is 1, so the moment generating function of a normal random variable is

$$M_X(t) = e^{\mu t + \sigma^2 t^2/2}$$

Differentiating this function twice with respect to t and setting $t = 0$ in the result, we find

$$\left.\frac{dM_X(t)}{dt}\right|_{t=0} = \mu_1' = \mu$$

$$\left.\frac{d^2M_X(t)}{dt^2}\right|_{t=0} = \mu_2' = \sigma^2 + \mu^2$$

Therefore, the variance of the normal random variable is

$$\sigma^2 = \mu_2' - \mu^2 = \sigma^2 + \mu^2 - \mu^2 = \sigma^2$$

Moment generating functions have many important and useful properties. One of the most important of these is the **uniqueness property.** That is, the moment generating function of a random variable is unique when it exists, so that if we have two random variables X and Y, say, with moment generating functions $M_X(t)$ and $M_Y(t)$, then if $M_X(t) = M_Y(t)$ for all values of t, both X and Y have the same probability distribution. Some of the other useful properties of the moment generating function are summarized on the following page.

Theorem II-2

If X is a random variable and a is a constant, then

$$M_{X+a}(t) = e^{at}M_X(t)$$

Proof

$$\begin{aligned} M_{X+a}(t) &= E[e^{t(X+a)}] \\ &= e^{at}E(e^{tX}) \\ &= e^{at}M_X(t) \end{aligned}$$

Theorem II-3

If X is a random variable and a is a constant, then

$$M_{aX}(t) = M_X(at)$$

Proof

$$\begin{aligned} M_{aX}(t) &= E[e^{t(aX)}] \\ &= E[e^{(at)X}] \\ &= M_X(at) \end{aligned}$$

Theorem II-4

If $X_1, X_2, \ldots, X_n$ are independent random variables with moment generating functions $M_{X_1}(t), M_{X_2}(t), \ldots, M_{X_n}(t)$, respectively, and if $Y = X_1 + X_2 + \cdots + X_n$, then the moment generating function of Y is

$$M_Y(t) = M_{X_1}(t) \cdot M_{X_2}(t) \cdot \cdots \cdot M_{X_n}(t)$$

Proof For the case where the X's are continuous random variables

$$M_Y(t) = E(e^{tY}) = E[e^{t(X_1+X_2+\cdots+X_n)}]$$
$$= \int_{-\infty}^{\infty}\int_{-\infty}^{\infty}\cdots\int_{-\infty}^{\infty} e^{t(x_1+x_2+\cdots+x_n)} f(x_1, x_2, \ldots, x_n)\, dx_1\, dx_2 \ldots dx_n$$

Since the X's are independent,

$$f(x_1, x_2, \ldots, x_n) = f_{X_1}(x_1)\cdot f_{X_2}(x_2)\cdot \cdots \cdot f_{X_n}(x_n)$$

and one may write

$$M_Y(t) = \int_{-\infty}^{\infty} e^{tx_1} f_{X_1}(x_1)\, dx_1 \int_{-\infty}^{\infty} e^{tx_2} f_{X_2}(x_2)\, dx_2 \cdots \int_{-\infty}^{\infty} e^{tx_n} f_{X_n}(x_n)\, dx_n$$
$$= M_{X_1}(t)\cdot M_{X_2}(t)\cdot \cdots \cdot M_{X_n}(t)$$

For the case when the X's are discrete, we would use the same approach replacing integrals with summations.

Theorem II-4 is particularly useful. In many situations we need to find the distribution of the sum of two or more independent random variables, and often this theorem makes the problem very easy. This is illustrated in the following example.

EXAMPLE II-3

Suppose that X_1 and X_2 are two independent Poisson random variables with parameters λ_1, and λ_2 respectively. Find the probability distribution of $Y = X_1 + X_2$.

The moment generating function of a Poisson random variable with parameter λ is

$$M_X(t) = e^{\lambda(e^t-1)}$$

so the moment generating functions of X_1 and X_2 are $M_{X_1}(t) = e^{\lambda_1(e^t-1)}$ and $M_{X_2}(t) = e^{\lambda_2(e^t-1)}$, respectively. Using the results of Theorem II-4, we find that the moment generating function of $Y = X_1 + X_2$ is

$$M_Y(t) = M_{X_1}(t)M_{X_2}(t)$$
$$= e^{\lambda_1(e^t-1)}e^{\lambda_2(e^t-1)}$$
$$= e^{(\lambda_1+\lambda_2)(e^t-1)}$$

which is recognized as the moment generating function of a Poisson random variable with parameter $\lambda_1 + \lambda_2$. Therefore, we have shown that the sum of two independent Poisson random variables with parameters λ_1 and λ_2 is a Poisson random variable with parameter equal to the sum of the two parameters $\lambda_1 + \lambda_2$.

EXERCISES

II-1. A random variable X has the discrete uniform distribution

$$f(x) = \begin{cases} \frac{1}{m}, & x = 1, 2, \ldots, m \\ 0, & \text{otherwise} \end{cases}$$

(a) Show that the moment generating function is

$$M_X(t) = \frac{e^t(1 - e^{tm})}{m(1 - e^t)}$$

(b) Use $M_X(t)$ to find the mean and variance of X.

II-2. A random variable X has the Poisson distribution

$$f(x) = \begin{cases} \frac{e^{-\lambda}\lambda^x}{x!}, & x = 0, 1, \ldots \\ 0, & \text{otherwise} \end{cases}$$

(a) Show that the moment generating function is

$$M_X(t) = e^{\lambda(e^t - 1)}$$

(b) Use $M_X(t)$ to find the mean and variance of the Poisson random variable.

II-3. The geometric random variable X has probability distribution

$$f(x) = \begin{cases} (1 - p)^{x-1}p, & x = 1, 2, \ldots \\ 0, & \text{otherwise} \end{cases}$$

(a) Show that the moment generating function is

$$M_X(t) = \frac{pe^t}{1 - (1 - p)e^t}$$

(b) Use $M_X(t)$ to find the mean and variance of X.

II-4. The chi-square random variable with k degrees of freedom has moment generating function $M_X(t) = (1 - 2t)^{-k/2}$. Suppose that X_1 and X_2 are independent chi-square random variables with k_1 and k_2 degrees of freedom, respectively. What is the distribution of $Y = X_1 + X_2$?

II-5. A continuous random variable X has the following probability distribution:

$$f(x) = \begin{cases} 4xe^{-2x}, & x > 0 \\ 0, & \text{otherwise} \end{cases}$$

(a) Find the moment generating function for X.

(b) Find the mean and variance of X.

II-6. The continuous uniform random variable X has density function

$$f(x) = \begin{cases} \frac{1}{\beta - \alpha}, & \alpha \le x \le \beta \\ 0, & \text{otherwise} \end{cases}$$

(a) Show that the moment generating function is

$$M_X(t) = \frac{e^{t\beta} - e^{t\alpha}}{t(\beta - \alpha)}$$

(b) Use $M_X(t)$ to find the mean and variance of X.

II-7. A random variable X has the exponential distribution

$$f(x) = \begin{cases} \lambda e^{-\lambda x}, & x > 0 \\ 0, & \text{otherwise} \end{cases}$$

(a) Show that the moment generating function of X is

$$M_X(t) = \left(1 - \frac{t}{\lambda}\right)^{-1}$$

(b) Find the mean and variance of X.

II-8. A random variable X has the gamma distribution

$$f(x) = \begin{cases} \frac{\lambda}{\Gamma(r)}(\lambda x)^{r-1}e^{-\lambda x}, & x > 0 \\ 0, & \text{otherwise} \end{cases}$$

(a) Show that the moment generating function of X is

$$M_X(t) = \left(1 - \frac{t}{\lambda}\right)^{-r}$$

(b) Find the mean and variance of X.

II-9. Let $X_1, X_2, \ldots, X_r$ be independent exponential random variables with parameter λ.
(a) Find the moment generating function of $Y = X_1 + X_2 + \cdots + X_r$.
(b) What is the distribution of the random variable Y?
Hint: Use the results of Exercise II-8.

II-10. Suppose that X_i has a normal distribution with mean μ_i and variance σ_i^2, $i = 1, 2$. Let X_1 and X_2 be independent.
(a) Find the moment generating function of $Y = X_1 + X_2$.
(b) What is the distribution of the random variable Y?

II-11. Show that the moment generating function of the chi-square random variable with k degrees of freedom is $M_X(t) = (1 - 2t)^{-k/2}$. Show that the mean and variance of this random variable are k and $2k$, respectively.

II-12. (a) Show that by expanding e^{tX} in a power series and taking expectations term by term we may write the moment generating function as

$$\begin{aligned} M_X(t) &= E(e^{tX}) \\ &= 1 + \mu_1' t + \mu_2' \frac{t^2}{2!} + \cdots \\ &\quad + \mu_r' \frac{t^r}{r!} + \cdots \end{aligned}$$

Thus, the coefficient of $t^r/r!$ in this expansion is μ_r', the rth origin moment.
(b) Consider the gamma random variable in Exercise II-8 above. Write the power series expansion for $M_X(t)$.
(c) Find μ_1' and μ_2' using the results of parts (a) and (b). Does this approach give the same answers that you found for the mean and variance of the gamma random variable in Exercise II-8?

III. FUNCTIONS OF RANDOM VARIABLES

In many situations in statistics, it is necessary to derive the probability distribution of a function of one or more random variables. In this section, we present some results that are helpful in solving this problem.

Suppose that X is a discrete random variable with probability distribution $f_X(x)$. Let $Y = h(X)$ be a function of X that defines a one-to-one transformation between the values of X and Y, and we wish to find the probability distribution of Y. By a one-to-one transformation, we mean that each value x is related to one and only one value of $y = h(x)$, and that each value of y is related to one and only one value of x, say, $x = u(y)$, where $u(y)$ is found by solving $y = h(x)$ for x in terms of y.

Now, the random variable Y takes on the value y when X takes on the value $u(y)$. Therefore, the probability distribution of Y is

$$f_Y(y) = P(Y = y) = P[X = u(y)] = f_X[u(y)]$$

We may state this result as a theorem.

Theorem III-1

Suppose that X is a **discrete** random variable with probability distribution $f_X(x)$. Let $Y = h(X)$ define a one-to-one transformation between the values of X and Y so that the equation $y = h(x)$ can be solved uniquely for x in terms of y. Let this solution be $x = u(y)$. Then the probability distribution of the random variable Y is

$$f_Y(y) = f_X[u(y)]$$

EXAMPLE III-1

Let X be a geometric random variable with probability distribution

$$f_X(x) = \begin{cases} p(1-p)^{x-1}, & x = 1, 2, \ldots \\ 0, & \text{otherwise} \end{cases}$$

Find the probability distribution of $Y = X^2$.

Since $X \geq 0$ the transformation is one to one; that is, $y = x^2$ and $x = \sqrt{y}$. Therefore, the theorem indicates that the distribution of the random variable Y is

$$f_Y(y) = \begin{cases} f(\sqrt{y}) = p(1-p)^{\sqrt{y}-1}, & y = 1, 4, 9, 16, \ldots \\ 0, & \text{otherwise} \end{cases}$$

Now suppose that we have two discrete random variables X_1 and X_2 with joint probability distribution $f_{X_1X_2}(x_1, x_2)$ and we wish to find the joint probability distribution $f_{Y_1Y_2}(y_1, y_2)$ of two new random variables $Y_1 = h_1(X_1, X_2)$ and $Y_2 = h_2(X_1, X_2)$. We assume that the functions h_1 and h_2 define a one-to-one transformation between (x_1, x_2) and (y_1, y_2). Solving the equations $y_1 = h_1(x_1, x_2)$ and $y_2 = h_2(x_1, x_2)$ simultaneously, we obtain the unique solution $x_1 = u_1(y_1, y_2)$ and $x_2 = u_2(y_1, y_2)$. Therefore, the random variables Y_1 and Y_2 take on the values y_1 and y_2 when X_1 takes on the value $u_1(y_1, y_2)$ and X_2 takes the value $u_2(y_1, y_2)$. The joint probability distribution of Y_1 and Y_2 is

$$\begin{aligned} f_{Y_1Y_2}(y_1, y_2) &= P(Y_1 = y_1, Y_2 = y_2) \\ &= P[X_1 = u_1(y_1, y_2), X_2 = u_2(y_1, y_2)] \\ &= f_{X_1X_2}[u_1(y_1, y_2), u_2(y_1, y_2)] \end{aligned}$$

We will also state this result as a theorem.

Theorem III-2

Suppose that X_1 and X_2 are **discrete** random variables with joint probability distribution $f_{X_1X_2}(x_1, x_2)$, and let $Y_1 = h_1(X_1, X_2)$ and $Y_2 = h_2(X_1, X_2)$ define one-to-one transformations between the points (x_1, x_2) and (y_1, y_2), so that the equations $y_1 = h_1(x_1, x_2)$ and $y_2 = h_2(x_1, x_2)$ can be solved uniquely for x_1 and x_2 in terms of y_1 and y_2. Let this solution be $x_1 = u_1(y_1, y_2)$ and $x_2 = u_2(y_1, y_2)$. Then the joint probability distribution of Y_1 and Y_2 is

$$f_{Y_1Y_2}(y_1, y_2) = f_{X_1X_2}[u_1(y_1, y_2), u_2(y_1, y_2)]$$

A very important application of Theorem III-2 is in finding the distribution of a random variable Y_1 that is a function of two other random variables X_1 and X_2. That is, let $Y_1 = h_1(X_1, X_2)$ where X_1 and X_2 are discrete random variables with joint distribution $f_{X_1X_2}(x_1, x_2)$. We want to find the probability distribution of Y_1, say, $f_{Y_1}(y_1)$. To do this, we define a second function $Y_2 = h_2(X_1, X_2)$, so that the one-to-one correspondence between the points (x_1, x_2) and (y_1, y_2) is maintained, and we use the result in Theorem III-2 to find the joint probability distribution of Y_1 and Y_2. Then the distribution of Y_1 alone is found by summing over the y_2 values in this joint distribution. That is, $f_{Y_1}(y_1)$ is just the **marginal probability distribution** of Y_1, or

$$f_{Y_1}(y_1) = \sum_{y_2} f_{Y_1Y_2}(y_1, y_2)$$

EXAMPLE III-2

Consider the case where X_1 and X_2 are independent Poisson random variables with parameters λ_1 and λ_2, respectively. We will find the distribution of the random variable $Y_1 = X_1 + X_2$.

The joint distribution of X_1 and X_2 is

$$\begin{aligned} f_{X_1X_2}(x_1, x_2) &= f_{X_1}(x_1) f_{X_2}(x_2) \\ &= \frac{e^{-\lambda_1}\lambda_1^{x_1}}{x_1!} \frac{e^{-\lambda_2}\lambda_2^{x_2}}{x_2!} \\ &= \frac{e^{-(\lambda_1+\lambda_2)}\lambda_1^{x_1}\lambda_2^{x_2}}{x_1!\,x_2!}, \qquad x_1 = 0, 1, \ldots, \qquad x_2 = 0, 1, \ldots \end{aligned}$$

because X_1 and X_2 are independent. Now to use Theorem III-2 we need to define a second function $Y_2 = h_2(X_1, X_2)$. Let this function be $Y_2 = X_2$. Now the solutions for x_1 and x_2 are $x_1 = y_1 - y_2$ and $x_2 = y_2$. Thus, from Theorem III-2 the joint probability distribution of Y_1 and Y_2 is

$$f_{Y_1Y_2}(y_1, y_2) = \frac{e^{-(\lambda_1+\lambda_2)}\lambda_1^{(y_1-y_2)}\lambda_2^{y_2}}{(y_1 - y_2)!\, y_2!}, \qquad y_1 = 0, 1, 2, \ldots, \qquad y_2 = 0, 1, \ldots, y_1$$

Because $x_1 \geq 0$, the transformation $x_1 = y_1 - y_2$ requires that $x_2 = y_2$ must always be less than or equal to y_1. Thus, the values of y_2 are $0, 1, \ldots, y_1$, and the marginal probability distribution of Y_1 is obtained as follows:

$$f_{Y_1}(y_1) = \sum_{y_2=0}^{y_1} f_{Y_1Y_2}(y_1, y_2) = \sum_{y_2=0}^{y_1} \frac{e^{-(\lambda_1+\lambda_2)}\lambda_1^{(y_1-y_2)}\lambda_2^{y_2}}{(y_1 - y_2)!\, y_2!}$$

$$= \frac{e^{-(\lambda_1+\lambda_2)}}{y_1!} \sum_{y_2=0}^{y_1} \frac{y_1!\, \lambda_1^{(y_1-y_2)}\lambda_2^{y_2}}{(y_1 - y_2)!\, y_2!} = \frac{e^{-(\lambda_1+\lambda_2)}}{y_1!} \sum_{y_2=0}^{y_1} \binom{y_1}{y_2} \lambda_1^{(y_1-y_2)}\lambda_2^{y_2}$$

The summation in this last expression is the binomial expansion of $(\lambda_1 + \lambda_2)^{y_1}$, so

$$f_{Y_1}(y_1) = \frac{e^{-(\lambda_1+\lambda_2)}(\lambda_1 + \lambda_2)^{y_1}}{y_1!}, \qquad y_1 = 0, 1, \ldots$$

We recognize this as a Poisson distribution with parameter $\lambda_1 + \lambda_2$. Therefore, we have shown that the sum of two independent Poisson random variables with parameters λ_1 and λ_2 has a Poisson distribution with parameter $\lambda_1 + \lambda_2$.

We now consider the situation where the random variables are continuous. Theorem III-3 treats the case where $Y = h(X)$, with X continuous and the transformation is one to one.

Theorem III-3

Suppose that X is a **continuous** random variable with probability distribution $f_X(x)$. The function $Y = h(X)$ is a one-to-one transformation between the values of Y and X so that the equation $y = h(x)$ can be uniquely solved for x in terms of y. Let this solution be $x = u(y)$. The probability distribution of Y is

$$f_Y(y) = f_X[u(y)]\,|J|$$

where $J = u'(y)$ is called the **Jacobian** of the transformation and the absolute value of J is used.

Proof Let the function $y = h(x)$ be an increasing function of x. Now

$$\begin{aligned} P(Y \le a) &= P[X \le u(a)] \\ &= \int_{-\infty}^{u(a)} f_X(x)\,dx \end{aligned}$$

If we change the variable of integration from x to y by using $x = u(y)$, we obtain $dx = u'(y)\,dy$ and then

$$P(Y \le a) = \int_{-\infty}^{a} f_X[u(y)]u'(y)\,dy$$

Since the integral gives the probability that $Y \le a$ for all values of a contained in the feasible set of values for y, then $f_X[u(y)]u'(y)$ must be the probability distribution of Y. Therefore, the probability distribution of Y is

$$f_Y(y) = f_X[u(y)]u'(y) = f_X[u(y)]\,J$$

If the function $y = h(x)$ is a decreasing function of x, a similar argument holds.

EXAMPLE III-3

Let X be a continuous random variable with probability distribution

$$f_X(x) = \begin{cases} \dfrac{x}{8}, & 0 \le x < 4 \\ 0, & \text{otherwise} \end{cases}$$

Find the probability distribution of $Y = h(X) = 2X + 4$.

Note that $y = h(x) = 2x + 4$ is an increasing function of x. The inverse solution is $x = u(y) = (y - 4)/2$, and from this we find the Jacobian to be $J = u'(y) = dx/dy = 1/2$. Therefore, from the theorem the probability distribution of Y is

$$f_Y(y) = \begin{cases} \dfrac{(y-4)/2}{8}\left(\dfrac{1}{2}\right) = \dfrac{y-4}{32}, & 4 \le y \le 12 \\ 0, & \text{otherwise} \end{cases}$$

We now consider the case where X_1 and X_2 are continuous random variables and we wish to find the joint probability distribution of $Y_1 = h_1(X_1, X_2)$ and $Y_2 = h_2(X_1, X_2)$ where the transformation is one-to-one. The application of this will typically be in finding the probability distribution of $Y_1 = h_1(X_1, X_2)$, analogous to the discrete case discussed above. We will need the following theorem.

Theorem III-4

Suppose that X_1 and X_2 are **continuous** random variables with joint probability distribution $f_{X_1X_2}(x_1, x_2)$, and let $Y_1 = h_1(X_1, X_2)$ and $Y_2 = h_2(X_1, X_2)$ define a one-to-one transformation between the points (x_1, x_2) and (y_1, y_2). Let the equations $y_1 = h_1(x_1, x_2)$ and $y_2 = h_2(x_1, x_2)$ be uniquely solved for x_1 and x_2 in terms of y_1 and y_2 as $x_1 = u_1(y_1, y_2)$ and $x_2 = u_2(y_1, y_2)$. Then the joint probability of Y_1 and Y_2 is

$$f_{Y_1Y_2}(y_1, y_2) = f_{X_1X_2}[u_1(y_1, y_2), u_2(y_1, y_2)]\,|J|$$

where J is the *Jacobian,* and is given by the following determinant:

$$J = \begin{vmatrix} \partial x_1/\partial y_1, & \partial x_1/\partial y_2 \\ \partial x_2/\partial y_1, & \partial x_2/\partial y_2 \end{vmatrix}$$

and the absolute value of the determinant is used.

This theorem can be used to find $f_{Y_1Y_2}(y_1, y_2)$, the joint probability distribution of Y_1 and Y_2. Then the probability distribution of Y_1 is

$$f_{Y_1}(y_1) = \int_{-\infty}^{\infty} f_{Y_1Y_2}(y_1, y_2)\, dy_2$$

That is, $f_{Y_1}(y_1)$ is the **marginal probability distribution** of Y_1.

EXAMPLE III-4

Suppose that X_1 and X_2 are independent exponential random variables with $f_{X_1}(x_1) = 2e^{-2x_1}$ and $f_{X_2}(x_2) = 2e^{-2x_2}$. Find the probability distribution of $Y = X_1/X_2$.

The joint probability distribution of X_1 and X_2 is

$$f_{X_1X_2}(x_1, x_2) = \begin{cases} 4e^{-2(x_1+x_2)}, & x_1 \geq 0, \quad x_2 \geq 0 \\ 0, & \text{otherwise} \end{cases}$$

because X_1 and X_2 are independent. Let $Y_1 = h_1(X_1, X_2) = X_1/X_2$ and $Y_2 = h_2(X_1, X_2) = X_1 + X_2$. The inverse solutions of $y_1 = x_1/x_2$ and $y_2 = x_1 + x_2$ are $x_1 = y_1y_2/(1 + y_1)$ and $x_2 = y_2/(1 + y_1)$, and it follows that

$$\frac{\partial x_1}{\partial y_1} = y_2\left[\frac{1}{(1 + y_1)^2}\right], \qquad \frac{\partial x_1}{\partial y_2} = \left[\frac{y_1}{(1 + y_1)}\right]$$

$$\frac{\partial x_2}{\partial y_1} = y_2\left[\frac{-1}{(1 + y_1)^2}\right], \qquad \frac{\partial x_2}{\partial y_2} = \left[\frac{1}{(1 + y_1)}\right]$$

Therefore

$$J = \begin{vmatrix} \dfrac{y_2}{(1+y_1)^2} & \dfrac{y_1}{(1+y_1)} \\ \dfrac{-y_2}{(1+y_1)^2} & \dfrac{1}{(1+y_1)} \end{vmatrix} = \frac{y_2}{(1+y_1)^2}$$

and from Theorem III-4 the joint probability distribution of Y_1 and Y_2 is

$$\begin{aligned} f_{Y_1Y_2}(y_1, y_2) &= f_{X_1X_2}[u_1(y_1, y_2), u_2(y_1, y_2)]\,|J| \\ &= 4e^{-2[y_1y_2/(1+y_1)+y_2/(1+y_1)]} \left|\frac{y_2}{(1+y_1)^2}\right| \\ &= 4e^{-2y_2}\, y_2/(1+y_1)^2 \end{aligned}$$

for $y_1 > 0$, $y_2 > 0$. We need to find the distribution of $Y_1 = X_1/X_2$. This is the marginal probability distribution of Y_1, or

$$\begin{aligned} f_{Y_1}(y_1) &= \int_0^\infty f_{Y_1Y_2}(y_1, y_2)\, dy_2 \\ &= \int_0^\infty 4e^{-2y_2}\,[y_2/(1+y_1)^2]\, dy_2 \\ &= \begin{cases} \dfrac{1}{(1+y_1)^2}, & y_1 > 0 \\ 0, & \text{otherwise} \end{cases} \end{aligned}$$

In some problems involving transformations, we need to find the probability distribution of the random variable $Y = h(X)$ when X is a continuous random variable, but the transformation is not one-to-one. The following theorem is helpful.

Theorem III-5

Suppose that X is a **continuous** random variable with probability distribution $f_X(x)$, and $Y = h(X)$ is a transformation that is not one-to-one. If the interval over which X is defined can be partitioned into m mutually exclusive disjoint sets such that each of the inverse functions $x_1 = u_1(y)$, $x_2 = u_2(y)$, ..., $x_m = u_m(y)$ of $y = u(x)$ is one-to-one, then the probability distribution of Y is

$$f_Y(y) = \sum_{i=1}^{m} f_X[u_i(y)]\,|J_i|$$

where $J_i = u_i'(y)$, $i = 1, 2, \ldots, m$ and the absolute values are used.

To illustrate how this theorem is used, suppose that X is a normal random variable with mean μ and variance σ^2, and we wish to show that the distribution of $Y = (X - \mu)^2/\sigma^2$ is a chi-square distribution with 1 degree of freedom. Let $Z = (X - \mu)/\sigma$, and $Y = Z^2$. The probability distribution of Z is the standard normal; that is,

$$f(z) = \frac{1}{\sqrt{2\pi}} e^{-z^2/2}, \qquad -\infty < z < \infty$$

The inverse solutions of $y = z^2$ are $z = \pm\sqrt{y}$, so the transformation is not one-to-one. Define $z_1 = -\sqrt{y}$ and $z_2 = +\sqrt{y}$ so that $J_1 = -(1/2)/\sqrt{y}$ and $J_2 = (1/2)/\sqrt{y}$. Then by Theorem III-5, the probability distribution of Y is

$$\begin{aligned} f_Y(y) &= \frac{1}{\sqrt{2\pi}} e^{-y/2}\left|\frac{-1}{2\sqrt{y}}\right| + \frac{1}{\sqrt{2\pi}} e^{-y/2}\left|\frac{1}{2\sqrt{y}}\right| \\ &= \frac{1}{2^{1/2}\sqrt{\pi}} y^{1/2-1} e^{-y/2}, \qquad y > 0 \end{aligned}$$

Now it can be shown that $\sqrt{\pi} = \Gamma(1/2)$, so we may write $f(y)$ as

$$f_Y(y) = \begin{cases} \dfrac{1}{2^{1/2}\Gamma\left(\dfrac{1}{2}\right)} y^{1/2-1} e^{-y/2}, & y > 0 \\ 0, & \text{otherwise} \end{cases}$$

which is the chi-squared distribution with 1 degree of freedom.

EXERCISES

III-1. Suppose that X is a random variable with probability distribution

$$f_X(x) = \begin{cases} 1/4, & x = 1, 2, 3, 4 \\ 0, & \text{otherwise} \end{cases}$$

Find the probability distribution of the random $Y = 2X + 1$.

III-2. Let X be a binomial random variable with $p = 0.25$ and $n = 3$. Find the probability distribution of the random variable $Y = X^2$.

III-3. Suppose that X is a continuous random variable with probability distribution

$$f_X(x) = \begin{cases} \dfrac{x}{18}, & 0 \le x \le 6 \\ 0, & \text{otherwise} \end{cases}$$

(a) Find the probability distribution of the random variable $Y = 2X + 10$.

(b) Find the expected value of Y.

III-4. Suppose that X has a uniform probability distribution

$$f_X(x) = \begin{cases} 1, & 0 \le x \le 1 \\ 0, & \text{otherwise} \end{cases}$$

Show that the probability distribution of the random variable $Y = -2 \ln X$ is chi-squared with two degrees of freedom.

III-5. A current of I amperes flows through a resistance of R ohms according to the probability distribution

$$f_I(i) = \begin{cases} 2i, & 0 \le i \le 1 \\ 0, & \text{otherwise} \end{cases}$$

Suppose that the resistance is also a random variable with probability distribution

$$f_R(r) = \begin{cases} 1, & 0 \le r \le 1 \\ 0, & \text{otherwise} \end{cases}$$

Assume that I and R are independent.

(a) Find the probability distribution for the power (in watts) $P = I^2R$.

(b) Find $E(P)$.

III-6. A random variable X has the following probability distribution:

$$f_X(x) = \begin{cases} e^{-x}, & x \ge 0 \\ 0, & \text{otherwise} \end{cases}$$

(a) Find the probability distribution for $Y = X^2$.

(b) Find the probability distribution for $Y = X^{1/2}$.

(c) Find the probability distribution for $Y = \ln X$.

III-7. The velocity of a particle in a gas is a random variable V with probability distribution

$$f_V(v) = \begin{cases} av^2e^{-bv} & v > 0 \\ 0, & \text{otherwise} \end{cases}$$

where b is a constant that depends on the temperature of the gas and the mass of the particle.

(a) Find the value of the constant a.

(b) The kinetic energy of the particle is $W = mV^2/2$. Find the probability distribution of W.

III-8. Suppose that X has the probability distribution

$$f_X(x) = \begin{cases} 1, & 1 \le x \le 2 \\ 0, & \text{otherwise} \end{cases}$$

Find the probability distribution of the random variable $Y = e^X$.

III-9. Prove that Theorem III-3 holds when $y = h(x)$ is a decreasing function of x.

III-10. The random variable X has the probability distribution

$$f_X(x) = \begin{cases} \frac{x}{8}, & 0 \le x \le 4 \\ 0, & \text{otherwise} \end{cases}$$

Find the probability distribution of $Y = (X - 2)^2$.

III-11. Consider a rectangle with sides of length S_1 and S_2, where S_1 and S_2 are independent random variables. The probability distributions of S_1 and S_2 are

$$f_{S_1}(s_1) = \begin{cases} 2s_1, & 0 \le s_1 \le 1 \\ 0, & \text{otherwise} \end{cases}$$

and

$$f_{S_2}(s_2) = \begin{cases} \frac{s_2}{8}, & 0 \le s_2 \le 4 \\ 0, & \text{otherwise} \end{cases}$$

(a) Find the joint distribution of the area of the rectangle $A = S_1S_2$ and the random variable $Y = S_1$.

(b) Find the probability distribution of the area A of the rectangle.

III-12. Suppose we have a simple electrical circuit in which Ohm's law $V = IR$ holds. We are interested in the probability distribution of the resistance R given that V and I are independent random variables with the following distributions:

$$f_V(v) = \begin{cases} e^{-v}, & v \ge 0 \\ 0, & \text{otherwise} \end{cases}$$

and

$$f_I(i) = \begin{cases} 1, & 1 \le i \le 2 \\ 0, & \text{otherwise} \end{cases}$$

Find the probability distribution of R.

IV. DEVELOPMENT OF THE *t*- AND *F*-DISTRIBUTIONS

Throughout this book, we have made extensive use of **sampling distributions.** Two of these sampling distributions, the *t*- and the *F*-distributions, play a central role in statistical estimation and hypothesis testing. We now present the derivations that lead to these sampling distributions. This presentation makes use of the techniques for transformation of variables in Section III. We state the results as theorems.

Theorem IV-1. The *t*-Distribution

Let Z be a standard normal random variable and V be a chi-square random variable with k degrees of freedom. If Z and V are independent, then the distribution of the random variable

$$T = \frac{Z}{\sqrt{V/k}}$$

is the t-distribution with k degrees of freedom. The probability density function is

$$f(t) = \frac{\Gamma[(k+1)/2]}{\sqrt{\pi k}\,\Gamma(k/2)} \frac{1}{[(t^2/k)+1]^{(k+1)/2}}, \quad -\infty < t < \infty$$

Proof Since Z and V are independent, their probability distribution is

$$f_{ZV}(z, v) = \frac{v^{(k/2)-1}}{\sqrt{2\pi}\, 2^{k/2}\Gamma\left(\frac{k}{2}\right)} e^{-(z^2+v)/2}, \qquad -\infty < z < \infty, 0 < v < \infty$$

Using Theorem III-4, we define a new random variable $U = V$. Thus, the inverse solutions of

$$t = \frac{z}{\sqrt{v/k}}$$

and

$$u = v$$

are

$$z = t\sqrt{\frac{u}{k}}$$

and

$$v = u$$

The Jacobian is

$$J = \begin{vmatrix} \sqrt{\dfrac{u}{k}} & \dfrac{t}{2uk} \\ 0 & 1 \end{vmatrix} = \sqrt{\frac{u}{k}}$$

Thus,

$$|J| = \sqrt{\frac{u}{k}}$$

and the joint probability distribution of T and U is

$$f_{TU}(t, u) = \frac{\sqrt{u}}{\sqrt{2\pi k}\, 2^{k/2}\Gamma\left(\dfrac{k}{2}\right)} u^{(k/2)-1} e^{-[(u/k)t^2+u]/2}$$

Now, since $V > 0$, we must require that $U > 0$, and since $-\infty < Z < \infty$, then $-\infty < T < \infty$. On rearranging this last equation, we have

$$f_{TU}(t, u) = \frac{1}{\sqrt{2\pi k}\, 2^{k/2}\Gamma\left(\dfrac{k}{2}\right)} u^{(k-1)/2} e^{-(u/2)[(t^2/k)+1]}, \qquad 0 < u < \infty,\ -\infty < t < \infty$$

The probability distribution of T is found by

$$\begin{aligned} f_T(t) &= \int_0^\infty f_{TU}(t, u)\, du \\ &= \frac{1}{\sqrt{2\pi k}\, 2^{k/2}\Gamma\left(\dfrac{k}{2}\right)} \int_0^\infty u^{(k-1)/2} e^{-(u/2)[(t^2/k)+1]}\, du \\ &= \frac{\Gamma[(k+1)/2]}{\sqrt{\pi k}\,\Gamma\left(\dfrac{k}{2}\right)} \frac{1}{[(t^2/k)+1]^{(k+1)/2}}, \qquad -\infty < t < \infty \end{aligned}$$

This is the distribution given in Theorem IV-1.

The probability distribution of the random variable T was first published by W. S. Gosset in a famous 1908 paper. Gosset was employed by the Guiness Brewers in Ireland. Since his employer discouraged publication of employee research, Gosset published these results under the pseudonym "Student." As a result, this probability distribution is sometimes called the **Student t-distribution.**

Theorem IV-2. The F-Distribution

Let U_1 and U_2 be independent chi-squared random variables with ν_1 and ν_2 degrees of freedom, respectively. Then the ratio

$$F = \frac{U_1/\nu_1}{U_2/\nu_2}$$

has the probability density function

$$f(x) = \frac{\Gamma\left(\frac{\nu_1+\nu_2}{2}\right)\left(\frac{\nu_1}{\nu_2}\right)^{\nu_1/2} x^{\nu_1/2-1}}{\Gamma\left(\frac{\nu_1}{2}\right)\Gamma\left(\frac{\nu_2}{2}\right)\left[\left(\frac{\nu_1}{\nu_2}\right)x+1\right]^{(\nu_1+\nu_2)/2}}, \qquad 0<x<\infty$$

This is the F-distribution with ν_1 degrees of freedom in the numerator and ν_2 degrees of freedom in the denominator.

Proof Since U_1 and U_2 are independent chi-squared random variables their joint probability distribution is

$$f(u_1, u_2) = \frac{u_1^{\nu_1/2-1}\, u_2^{\nu_2/2-1}}{2^{\nu_1/2}\Gamma\left(\frac{\nu_1}{2}\right) 2^{\nu_2/2}\Gamma\left(\frac{\nu_2}{2}\right)} e^{-(u_1+u_2)/2}, \qquad 0<u_1, u_2<\infty$$

Using the method in Theorem III-4, define the new random variable $M = U_2$. The inverse solutions of

$$x = \left(\frac{u_1}{\nu_1}\right)\Big/\left(\frac{u_2}{\nu_2}\right) \quad \text{and} \quad m = u_2$$

are

$$u_1 = \frac{\nu_1}{\nu_2} mx \quad \text{and} \quad u_2 = m$$

Therefore, the Jacobian is

$$J = \begin{vmatrix} \dfrac{\nu_1}{\nu_2} m & \dfrac{\nu_1}{\nu_2} x \\ 0 & 1 \end{vmatrix} = \frac{\nu_1}{\nu_2} m$$

Thus, the joint probability density function of X and M is

$$f(x, m) = \frac{\left(\dfrac{\nu_1}{\nu_2} mx\right)^{\nu_1/2-1} m^{\nu_2/2-1} e^{-(1/2)[(\nu_1/\nu_2)mx+m]} \left(\dfrac{\nu_1}{\nu_2}\right) m}{2^{\nu_1/2}\Gamma\left(\dfrac{\nu_1}{2}\right) 2^{\nu_2/2}\Gamma\left(\dfrac{\nu_2}{2}\right)}, \qquad 0 < x, m < \infty$$

The probability density function of F is

$$\begin{aligned} f(x) &= \int_0^\infty f(x, m)\, dm \\ &= \frac{\left(\dfrac{\nu_1}{\nu_2} x\right)^{\nu_1/2-1} \left(\dfrac{\nu_1}{\nu_2}\right)}{2^{(\nu_1+\nu_2)/2}\Gamma\left(\dfrac{\nu_1}{2}\right)\Gamma\left(\dfrac{\nu_2}{2}\right)} \int_0^\infty m^{(\nu_1+\nu_2)/2-1} e^{-(m/2)[(\nu_1/\nu_2)x+1]}\, dm \end{aligned}$$

Substituting $z = \dfrac{m}{z}\left(\dfrac{\nu_1}{\nu_2}x + 1\right)$ and $dm = 2\left(\dfrac{\nu_1}{\nu_2}x + 1\right)^{-1} dz$, we obtain

$$\begin{aligned} f(x) &= \frac{\left(\dfrac{\nu_1}{\nu_2}\right)^{\nu_1/2} x^{\nu_1/2-1}}{2^{(\nu_1+\nu_2)/2}\Gamma\left(\dfrac{\nu_1}{2}\right)\Gamma\left(\dfrac{\nu_2}{2}\right)} \int_0^\infty \left(\frac{2z}{\dfrac{\nu_1}{\nu_2}x + 1}\right)^{(\nu_1+\nu_2)/2-1} e^{-z}\, 2\left(\frac{\nu_1}{\nu_2}x + 1\right)^{-1} dz \\ &= \frac{\left(\dfrac{\nu_1}{\nu_2}\right)^{\nu_1/2} x^{\nu_1/2-1}}{\Gamma\left(\dfrac{\nu_1}{2}\right)\Gamma\left(\dfrac{\nu_2}{2}\right)\left(\dfrac{\nu_1}{\nu_2}x + 1\right)^{(\nu_1+\nu_2)/2}} \int_0^\infty z^{(\nu_1+\nu_2)/2-1} e^{-z}\, dz \\ &= \frac{\Gamma\left(\dfrac{\nu_1+\nu_2}{2}\right)\left(\dfrac{\nu_1}{\nu_2}\right)^{\nu_1/2} x^{\nu_1/2-1}}{\Gamma\left(\dfrac{\nu_1}{2}\right)\Gamma\left(\dfrac{\nu_2}{2}\right)\left(\dfrac{\nu_1}{\nu_2}x + 1\right)^{(\nu_1+\nu_2)/2}}, \qquad 0 < x < \infty \end{aligned}$$

which is the probability density function in the theorem.

V. BAYESIAN APPROACH TO ESTIMATION

This book uses methods of statistical inference based on the information in the sample data. In effect, these methods interpret probabilities as relative frequencies. Sometimes we call probabilities that are interpreted in this manner **objective probabilities.** There is another approach to statistical inference, called the **Bayesian approach,** that combines sample information with other information that may be available prior to collecting the sample. In this section we briefly illustrate how this approach may be used in parameter estimation.

Suppose that the random variable X has a probability distribution that is a function of one parameter θ. We will write this probability distribution as $f(x|\theta)$. This notation implies that the exact form of the distribution of X is conditional on the value assigned to θ. The classical approach to estimation would consist of taking a random sample of size n from this distribution and then substituting the sample values x_i into the estimator for θ. This estimator could have been developed using the maximum likelihood approach, for example.

Suppose that we have some additional information about θ and that we can summarize that information in the form of a probability distribution for θ, say, $f(\theta)$. This probability distribution is often called the **prior distribution** for θ, and suppose that the mean of the prior is μ_0 and the variance is σ_0^2. This is a very novel concept insofar as the rest of this book is concerned because we are now viewing the parameter θ as a random variable. The probabilities associated with the prior distribution are often called **subjective probabilities,** in that they usually reflect the analyst's degree of belief regarding the true value of θ. The Bayesian approach to estimation uses the prior distribution for θ, $f(\theta)$, and the joint probability distribution of the sample, say, $f(x_1, x_2, \ldots, x_n|\theta)$ to find a **posterior distribution** for θ, say, $f(\theta|x_1, x_2, \ldots, x_n)$. This posterior distribution contains information both from the sample and the prior distribution for θ. In a sense, it expresses our degree of belief regarding the true value of θ after observing the sample data. It is easy conceptually to find the posterior distribution. The joint probability distribution of the sample X_1, $X_2, \ldots, X_n$ and the parameter θ (remember that θ is a random variable) is

$$f(x_1, x_2, \ldots, x_n, \theta) = f(x_1, x_2, \ldots, x_n|\theta)\, f(\theta)$$

and the marginal distribution of $X_1, X_2, \ldots, X_n$ is

$$f(x_1, x_2, \ldots, x_n) = \begin{cases} \sum_{\theta} f(x_1, x_2, \ldots, x_n, \theta), & \theta \text{ discrete} \\ \int_{-\infty}^{\infty} f(x_1, x_2, \ldots, x_n, \theta)\, d\theta, & \theta \text{ continuous} \end{cases}$$

Therefore, the desired distribution is

$$f(\theta|x_1, x_2, \ldots, x_n) = \frac{f(x_1, x_2 \ldots, x_n, \theta)}{f(x_1, x_2, \ldots, x_n)}$$

We define the **Bayes estimator** of θ as the value $\tilde{\theta}$ that corresponds to the mean of the posterior distribution $f(\theta|x_1, x_2, \ldots, x_n)$.

Sometimes, the mean of the posterior distribution of θ can be determined easily. As a function of θ, $f(\theta|x_1, \ldots, x_n)$ is a probability density function and $x_1, \ldots, x_n$ are just constants. Because θ enters into $f(\theta|x_1, \ldots, x_n)$ only through $f(x_1, \ldots, x_n, \theta)$ if $f(x_1, \ldots, x_n, \theta)$, as a function of θ, is recognized as a well-known probability function, then the posterior mean of θ can be deduced from the well-known distribution without integration or even calculation of $f(x_1, \ldots, x_n)$.

EXAMPLE V-1

Let $X_1, X_2, \ldots, X_n$ be a random sample from the normal distribution with mean μ and variance σ^2, where μ is unknown and σ^2 is known. Assume that the prior distribution for μ is normal with mean μ_0 and variance σ_0^2; that is

$$f(\mu) = \frac{1}{\sqrt{2\pi}\sigma_0} e^{-(\mu-\mu_0)^2/(2\sigma_0^2)} = \frac{1}{\sqrt{2\pi\sigma_0^2}} e^{-(\mu^2-2\mu_0\mu+\mu_0^2)/(2\sigma_0^2)}$$

The joint probability distribution of the sample is

$$f(x_1, x_2, \ldots, x_n|\mu) = \frac{1}{(2\pi\sigma^2)^{n/2}} e^{-(1/2\sigma^2)\sum_{i=1}^{n}(x_i-\mu)^2}$$
$$= \frac{1}{(2\pi\sigma^2)^{n/2}} e^{-(1/2\sigma^2)(\sum x_i^2 - 2\mu\sum x_i + n\mu^2)}$$

Thus, the joint probability distribution of the sample and μ is

$$f(x_1, x_2, \ldots, x_n, \mu)$$

$$= \frac{1}{(2\pi\sigma^2)^{n/2}\sqrt{2\pi}\sigma_0} e^{-(1/2)[(1/\sigma_0^2+n/\sigma^2)\mu^2-(2\mu_0/\sigma_0^2+2\sum x_i/\sigma^2)\mu+\sum x_i^2/\sigma^2+\mu_0^2/\sigma_0^2]}$$

$$= e^{-(1/2)\left[\left(\frac{1}{\sigma_0^2}+\frac{1}{\sigma^2/n}\right)\mu^2-2\left(\frac{\mu_0}{\sigma_0^2}+\frac{\bar{x}}{\sigma^2/n}\right)\mu\right]} h_1(x_1, \ldots, x_n, \sigma^2, \mu_0, \sigma_0^2)$$

Upon completing the square in the exponent

$$= e^{-(1/2)\left(\frac{1}{\sigma_0^2}+\frac{1}{\sigma^2/n}\right)\left[\mu^2-\left(\frac{(\sigma^2/n)\mu_0}{\sigma_0^2+\sigma^2/n}+\frac{\bar{x}\sigma_0^2}{\sigma_0^2+\sigma^2/n}\right)\right]^2} h_2(x_1, \ldots, x_n, \sigma^2, \mu_0, \sigma_0^2)$$

where $h_i(x_1, \ldots, x_n, \sigma^2, \mu_0, \sigma_0^2)$ is a function of the observed values, σ^2, μ_0, and σ_0^2.

Now, because $f(x_1, \ldots, x_n)$ does not depend on μ,

$$f(\mu|x_1, \ldots, x_n) = e^{-(1/2)\left(\frac{1}{\sigma_0^2}+\frac{1}{\sigma^2/n}\right)\left[\mu^2-\left(\frac{(\sigma^2/n)\mu_0+\sigma_0^2\bar{x}}{\sigma_0^2+\sigma^2/n}\right)\right]} h_3(x_1, \ldots, x_n, \sigma^2, \mu_0, \sigma_0^2)$$

This is recognized as a normal probability density function with posterior mean

$$\frac{\dfrac{\sigma^2}{n}\mu_0 + \sigma_0^2 \bar{x}}{\sigma_0^2 + \dfrac{\sigma^2}{n}}$$

and posterior variance

$$\left(\frac{1}{\sigma_0^2} + \frac{1}{\sigma^2/n}\right)^{-1} = \frac{\sigma_0^2(\sigma^2/n)}{\sigma_0^2 + \sigma^2/n}$$

Consequently, the Bayes estimate of μ is a weighted average of μ_0 and $\bar{x}$. For purposes of comparison, note that the maximum likelihood estimate of μ, is $\hat{\mu} = \bar{x}$.

To illustrate, suppose that we have a sample of size $n = 10$ from a normal distribution with unknown mean μ and variance $\sigma^2 = 4$. Assume that the prior distribution for μ is normal with mean $\mu_0 = 0$ and variance $\sigma_0^2 = 1$. If the sample mean is 0.75, then the Bayes estimate of μ is

$$\frac{\dfrac{4}{10}(0) + 1(0.75)}{1 + \dfrac{4}{10}} = \frac{0.75}{1.4} = 0.536$$

Note that the maximum likelihood estimate of μ is $\bar{x} = 0.75$.

There is a relationship between the Bayes estimator for a parameter and the maximum likelihood estimator of the same parameter. For large sample sizes, the two are nearly equivalent. In general, the difference between the two estimators is small compared to $1/\sqrt{n}$. In practical problems, a moderate sample size will produce approximately the same estimate by either the Bayes or maximum likelihood method, if the sample results are consistent with the assumed prior information. If the sample results are inconsistent with the prior assumptions, then the Bayes estimate may differ considerably from the maximum likelihood estimate. In these circumstances, if the sample results are accepted as being correct, the prior information must be incorrect. The maximum likelihood estimate would then be the better estimate to use.

If the sample results are very different from the prior information, the Bayes estimator will always tend to produce an estimate that is between the maximum likelihood estimate and the prior assumptions. If there is more inconsistency between the prior information and the sample, there will be more difference between the two estimates.

We may use Bayesian methods to construct interval estimates of parameters that are similar to confidence intervals. If the posterior density for θ has been obtained, we can construct an interval, usually centered at the posterior mean, that contains $100(1 - \alpha)$ percent of the posterior probability. Such an interval is called the $100(1 - \alpha)$ percent **Bayes interval** for the unknown parameter θ.

While in many cases the Bayes interval estimate for θ will be quite similar to a classical confidence interval with the same confidence coefficient, the interpretation of the two is very different. A confidence interval is an interval that, before the sample is taken, will include the unknown parameter θ with probability $1 - \alpha$. That is, the classical confidence interval relates to the relative frequency of an interval including θ. On the other hand, a Bayes interval is an interval that contains $100(1 - \alpha)$ percent of the posterior probability for θ. Since the posterior probability density measures a degree of belief about θ given the sample results, the Bayes interval provides a subjective degree of belief about θ rather than a frequency interpretation. The Bayes interval estimate of θ is affected by the sample results but is not completely determined by them.

EXAMPLE V-2

Suppose that the random variable X is normally distributed with mean μ and variance 4. The value of μ is unknown, but a reasonable prior distribution would be normal with mean 2 and variance 1. We can show that the posterior probability distribution for μ is a normal distribution with mean $[(n/4) + 1]^{-1}[(n\bar{x}/4) + 2]$ and variance $[(n/4) + 1]^{-1}$. Therefore, a $100(1 - \alpha)$ percent Bayes interval for μ that is symmetric around the posterior mean is

$$[(n/4) + 1]^{-1}[(n\bar{x}/4) + 2] - z_{\alpha/2}[(n/4) + 1]^{-1/2} \leq \mu \leq [(n/4) + 1]^{-1}[(n\bar{x}/4) + 2] + z_{\alpha/2}[(n/4) + 1]^{-1/2}$$

Now suppose that a random sample of size $n = 16$ is taken and the sample mean $\bar{x} = 2.5$. A 95% Bayes interval for μ would be

$$1.52 \leq \mu \leq 3.48$$

The Bayes interval is slightly shorter than the classical confidence interval, because using the prior information is equivalent to a slight increase in sample size.

EXERCISES

V-1. Suppose that X is a normal random variable with unknown mean μ and known variance σ^2. The prior distribution for μ is a normal distribution with mean μ_0 and variance σ_0^2.
(a) Find the posterior distribution for μ.
(b) Find the Bayes estimator for μ.

V-2. Suppose that X is a normal random variable with unknown mean μ and known variance σ^2. The prior distribution for μ is a uniform distribution defined over the interval $[a, b]$.
(a) Find the posterior distribution for μ.
(b) Find the Bayes estimator for μ.

V-3. Suppose that X is a Poisson random variable with parameter λ. Let the prior distribution for λ be a gamma distribution with parameters $m + 1$ and $(m + 1)/\lambda_0$.

(a) Find the posterior distribution for λ.
(b) Find the Bayes estimator for λ.

V-4. Suppose that X is a normal random variable with unknown mean and known variance $\sigma^2 = 9$. The prior distribution for μ is normal with $\mu_0 = 4$ and $\sigma_0^2 = 1$. A random sample of $n = 25$ observations is taken, and the sample mean is $\bar{x} = 4.85$.
(a) Find the Bayes estimate of μ.
(b) Set up a 95% Bayes interval estimate for μ.
(c) Compare the Bayes interval in part (b) with a classical 95% confidence interval for μ. Which interval is longer?

V-5. The weight of boxes of candy is a normal random variable with mean μ and variance 1/10 lb. The prior distribution for μ is normal with mean 5.03 lb. and variance 1/25 lb. A random sample of 10 boxes gives a sample mean of $\bar{x} = 5.05$ lb.
(a) Find the Bayes estimate of μ.
(b) Find a 95% Bayes interval for μ.
(c) If boxes that weigh less than 4.95 lb. are considered defective, what percentage of the boxes produced by this process is defective?

V-6. The time between failures of a machine has an exponential distribution with parameter λ. Suppose that the prior distribution for λ is exponential with mean 100 hr. Two machines are observed, and the average time between failures is $\bar{x} = 1125$ hr.
(a) Find the Bayes estimate for λ.
(b) What proportion of the machines do you think will fail before 1000 hr?

VI. LIKELIHOOD RATIO TESTS

Hypothesis testing is one of the most important techniques of statistical inference. Throughout this book we have presented many applications of hypothesis testing. While we have emphasized a heuristic development, many of these hypothesis-testing procedures can be developed using a general principle called the likelihood ratio principle. Tests developed by this method often turn out to be "best" test procedures in the sense that they minimize the type II error probability β among all tests that have the same type I error probability α.

The likelihood ratio principle is easy to illustrate. Suppose that the random variable X has a probability distribution that is described by an unknown parameter θ, say, $f(x, \theta)$. We wish to test the hypothesis H_0: θ is in Ω_0 versus H_1: θ is in Ω_1, where Ω_0 and Ω_1 are disjoint sets of values (such as H_0: $\mu \geq 0$ versus H_1: $\mu < 0$). Let $x_1, x_2, \ldots, x_n$ be the observations in a random sample. The joint distribution of these sample observations is

$$f(x_1, x_2, \ldots, x_n, \theta) = f(x_1, \theta) \cdot f(x_2, \theta) \cdot \cdots \cdot f(x_n, \theta)$$

Recall from our discussion of maximum likelihood estimation in Chapter 7 that the likelihood function, say $L(\theta)$, is just this joint distribution considered as a function of the parameter θ. The **likelihood ratio principle** for test construction consists of the following steps.

1. Find the largest value of the likelihood for any θ in Ω_0. This is done by finding the maximum likelihood estimator of θ restricted to values within Ω_0 and by substituting this value of θ back into the likelihood function. This results in a value of the likelihood function that we will call $L(\Omega_0)$.

2. Find the largest value of the likelihood for any θ in Ω_1. Call this the value of the likelihood function $L(\Omega_1)$.

3. Form the ratio

$$\lambda = \frac{L(\Omega_0)}{L(\Omega_1)}$$

This ratio λ is called the **likelihood ratio test statistic.**

The test procedure calls for rejecting the null hypothesis H_0 when the value of this ratio λ is small, say, whenever $\lambda < k$, where k is a constant. Thus, the likelihood ratio principle requires rejecting H_0 when $L(\Omega_1)$ is much larger than $L(\Omega_0)$, which would indicate that the same data are more compatible with the alternative hypothesis H_1 than with the null hypothesis H_0. Usually, the constant k would be selected to give a specified value for α, the type I error probability.

These ideas can be illustrated by a hypothesis-testing problem that we have studied before—that of testing whether the mean of a normal population has a specified value μ_0. Suppose that we have a sample of n observations from a normal population with unknown mean μ and unknown variance σ^2, say, $x_1, x_2, \ldots, x_n$. We wish to test the hypothesis H_0: $\mu = \mu_0$ versus H_1: $\mu \neq \mu_0$. The likelihood function of the sample is

$$L = \left(\frac{1}{\sigma\sqrt{2\pi}}\right)^n e^{-\sum_{i=1}^{n}(x_i-\mu)^2/(2\sigma^2)}$$

and the values of Ω_0 and Ω_1 are $\Omega_0 = \mu_0$ and $\Omega_1 = \{\mu: -\infty < \mu < \infty\}$, respectively. The values of μ and σ^2 that maximize L in Ω_1 are the usual maximum likelihood estimates for μ and σ^2:

$$\hat{\mu} = \frac{1}{n}\sum_{i=1}^{n} x_i = \bar{x}$$

$$\hat{\sigma}^2 = \frac{1}{n}\sum_{i=1}^{n}(x_i - \bar{x})^2$$

Substituting these values in L, we have

$$L(\Omega_1) = \left[\frac{1}{(2\pi/n)\sum(x_i - \bar{x})^2}\right]^{n/2} e^{-(n/2)}$$

To maximize L in Ω_0 we simply set $\mu = \mu_0$ and then find the value of σ^2 that maximizes L. This value is found to be

$$\hat{\sigma}^2 = \frac{1}{n}\sum_{i=1}^{n}(x_i - \mu_0)^2$$

which gives

$$L(\Omega_0) = \left[\frac{1}{(2\pi/n)\sum(x_i - \mu_0)^2}\right]^{n/2} e^{-(n/2)}$$

The likelihood ratio is

$$\lambda = \frac{L(\Omega_0)}{L(\Omega_1)} = \left[\frac{\sum(x_i - \bar{x})^2}{\sum(x_i - \mu_0)^2}\right]^{n/2}$$

Now since

$$\sum(x_i - \mu_0)^2 = \sum(x_i - \bar{x})^2 + n(\bar{x} - \mu_0)^2$$

we may write λ as

$$\lambda = \left\{\frac{1}{1 + \dfrac{n(\bar{x} - \mu_0)^2}{\sum(x_i - \bar{x})^2}}\right\}^{n/2}$$

$$= \left\{\frac{1}{1 + \left[\dfrac{n(\bar{x} - \mu_0)^2}{\sum(x_i - \bar{x})^2}\right]\left(\dfrac{n-1}{n-1}\right)}\right\}^{n/2}$$

$$= \left\{\frac{1}{1 + \left(\dfrac{1}{n-1}\right)\left[\dfrac{(\bar{x} - \mu_0)^2}{s^2/n}\right]}\right\}^{n/2}$$

Notice that $\left[\dfrac{(\bar{x} - \mu_0)^2}{s^2/n}\right] = t^2$ is the square of the value of a random variable that has the t-distribution with $n - 1$ degrees of freedom when the null hypothesis H_0: $\mu = \mu_0$ is true. So we may write the value of the likelihood ratio λ as

$$\lambda = \left\{\frac{1}{1 + [t^2/(n-1)]}\right\}^{n/2}$$

It is easy to find the value for the constant k that would lead to rejection of the null hypothesis H_0. Since we reject H_0 if $\lambda < k$, this implies that small values of λ support the alternative hypothesis. Clearly, λ will be small when t^2 is large. So instead of specifying k we can specify a constant c and reject H_0: $\mu = \mu_0$ if $t^2 > c$. The critical values of t would be the extreme values, either positive or negative, and if we wish to control the type I error probability at α, the critical region in terms of t would be

$$t < -t_{\alpha/2,n-1} \quad \text{and} \quad t > t_{\alpha/2,n-1}$$

or, equivalently, we would reject H_0: $\mu = \mu_0$ if $t^2 > c = t^2_{\alpha/2,n-1}$. Therefore, the likelihood ratio test for H_0: $\mu = \mu_0$ versus H_1: $\mu \neq \mu_0$ is the familiar single-sample t-test.

The procedure employed in this example to find the critical region for the likelihood ratio λ is used often. That is, typically, we can manipulate λ to produce a condition that is equivalent to $\lambda < k$, but one that is simpler to use.

The likelihood ratio principle is a very general procedure. Most of the tests presented in this book that utilize the t, chi-square, and F-distributions for testing means and variances of normal distributions are likelihood ratio tests. The principle can also be used in cases where the observations are dependent, or even in cases where their distributions are different. However, the likelihood function can be very complicated in some of these situations. To use the likelihood ratio principle we must specify the form of the distribution. Without such a specification, it is impossible to write the likelihood function, and so if we are unwilling to assume a particular probability distribution, the likelihood ratio principle cannot be used. This could lead to the use of the nonparametric test procedures discussed in Chapter 14.

VII. RANDOM FACTORS IN FACTORIAL EXPERIMENTS

Chapter 13 on factorial experiments focused primarily on the case where all the factors were fixed; that is, the experimenter specifically chose the levels, and the conclusions from the experiment are confined to those specific levels. Now we briefly consider the situation where one or more of the factors in a factorial experiment are random, using the two-factor factorial design as an illustration.

The Random-Effects Model

Suppose that we have two factors A and B arranged in a factorial experiment in which the levels of both factors are selected at random from larger populations of factor levels, and we wish to extend our conclusion to the entire population of factor levels. The observations are represented by the model

$$Y_{ijk} = \mu + \tau_i + \beta_j + (\tau\beta)_{ij} + \epsilon_{ijk} \begin{cases} i = 1, 2, \ldots, a \\ j = 1, 2, \ldots, b \\ k = 1, 2, \ldots, n \end{cases} \qquad \text{(VII-1)}$$

where parameters τ_i, β_j, $(\tau\beta)_{ij}$, and ϵ_{ijk}, are normally and independently distributed random variables with means zero and variances σ_τ^2, σ_β^2, $\sigma_{\tau\beta}^2$, and σ^2, respectively. As a result of these assumptions, the variance of any observation Y_{ijk} is

$$V(Y_{ijk}) = \sigma_\tau^2 + \sigma_\beta^2 + \sigma_{\tau\beta}^2 + \sigma^2$$

and σ_τ^2, σ_β^2, $\sigma_{\tau\beta}^2$, and σ^2 are called **variance components.** The hypotheses that we are interested in testing are H_0: $\sigma_\tau^2 = 0$, H_0: $\sigma_\beta^2 = 0$, and H_0: $\sigma_{\tau\beta}^2 = 0$. Notice the similarity to the single-factor experiment random-effects model discussed in Chapter 12.

The basic analysis of variance remains unchanged; that is, SS_A, SS_B, SS_{AB}, SS_T, and SS_E are all calculated as in the fixed-effects case. To construct the test statistics, we must examine the expected mean squares. They are

$$E(MS_A) = \sigma^2 + n\sigma_{\tau\beta}^2 + bn\sigma_\tau^2$$

$$E(MS_B) = \sigma^2 + n\sigma_{\tau\beta}^2 + an\sigma_\beta^2$$

$$E(MS_{AB}) = \sigma^2 + n\sigma_{\tau\beta}^2$$

and

$$E(MS_E) = \sigma^2 \qquad \text{(VII-2)}$$

Note from the expected mean squares that the appropriate statistic for testing the no-interaction hypothesis H_0: $\sigma_{\tau\beta}^2 = 0$ is

$$F_0 = \frac{MS_{AB}}{MS_E} \qquad \text{(VII-3)}$$

since if H_0 is true, both numerator and denominator of F_0 have expectation σ^2, and only if H_0 is false is $E(MS_{AB})$ greater than $E(MS_E)$. The ratio F_0 is distributed as $F_{(a-1)(b-1),ab(n-1)}$. Similarly, for testing that there is no main effect of factor A, or H_0: $\sigma_\tau^2 = 0$, we would use

$$F_0 = \frac{MS_A}{MS_{AB}} \qquad \text{(VII-4)}$$

which is distributed as $F_{a-1,(a-1)(b-1)}$, and for testing H_0: $\sigma_\beta^2 = 0$ the test statistic is

$$F_0 = \frac{MS_B}{MS_{AB}} \qquad \text{(VII-5)}$$

which is distributed as $F_{b-1,(a-1)(b-1)}$. These are all upper-tail, one-tail tests. Thus, the null hypotheses above would be rejected at the α level of significance if the calculated value of f_0 exceeds the upper α percentage point of the F-distribution. Notice that these test statistics are not the same as those used if both factors A and B are fixed. The expected mean squares are always used as a guide to test statistic construction.

The variance components may be estimated by equating the observed mean squares to their expected values and solving for the variance components. This yields

$$\hat{\sigma}^2 = MS_E$$

$$\hat{\sigma}_{\tau\beta}^2 = \frac{MS_{AB} - MS_E}{n}$$

$$\hat{\sigma}_\beta^2 = \frac{MS_B - MS_{AB}}{an} \qquad \text{(VII-6)}$$

$$\hat{\sigma}_\tau^2 = \frac{MS_A - MS_{AB}}{bn}$$

EXAMPLE VII-1

Two factors that may influence the breaking strength of cloth are being studied. Four test machines and three operators are chosen at random, and an experiment is run using cloth from the same production segment. The data are shown in Table VII-1, and the analysis of variance is in Table VII-2. Notice that the first four columns in Table VII -2 are computed as in a standard (fixed-effects model) analysis. The test statistics are computed according to equations (VII-3) through (VII-5). We will use $\alpha = 0.05$. The test statistic for the no-interaction hypothesis H_0: $\sigma_{\tau\beta}^2 = 0$ is

$$f_0 = \frac{MS_{AB}}{MS_E} = \frac{5.94}{3.75} = 1.584$$

Table VII-1 Breaking Strength Data for Example VII-1

	Machine			
Operator	1	2	3	4
A	113	113	111	113
	112	118	111	119
B	111	110	111	114
	112	111	109	112
C	109	112	114	111
	111	115	112	112

which would be compared to $f_{0.05,6,12} = 3.00$, and so H_0: $\sigma^2_{\tau\beta} = 0$ cannot be rejected. The P-value for this ratio is $P = 0.2338$. To test for no machine effect (H_0: $\sigma^2_{\beta} = 0$), we compute

$$f_0 = \frac{MS_B}{MS_{AB}} = \frac{8.11}{5.94} = 1.365$$

which would be compared to $f_{0.05,3,6} = 4.76$. The P-value for this ratio is $P = 0.3403$. Therefore, we conclude that machines do not significantly affect the breaking strength test results. To test for no operator effect (H_0: $\sigma^2_{\tau} = 0$), we compute

$$f_0 = \frac{MS_A}{MS_{AB}} = \frac{13.17}{5.94} = 2.217$$

which would be compared to $f_{0.05,2,6} = 5.14$. The P-value for this ratio is $P = 0.1904$. We conclude that the operators do not affect the breaking strength test results. The variance components may be estimated using equations VII-6 as follows:

Table VII-2 Analysis of Variance for Example VII-1

Source of Variation	Sum of Squares	Degrees of Freedom	Mean Square	f_0	P-value
Operators	26.33	2	13.17	2.217	0.1904
Machines	24.33	3	8.11	1.365	0.3403
Interaction	35.67	6	5.94	1.584	0.2338
Error	45.00	12	3.75		
Total	131.33	23			

$$\hat{\sigma}^2 = 3.75$$

$$\hat{\sigma}^2_{\tau\beta} = \frac{5.94 - 3.75}{2} = 1.10$$

$$\hat{\sigma}^2_{\tau} = \frac{13.17 - 5.94}{8} = 0.90$$

$$\hat{\sigma}^2_{\beta} = \frac{8.11 - 5.94}{6} = 0.36$$

All of the variance components $\hat{\sigma}^2_{\tau}$, $\hat{\sigma}^2_{\beta}$, and $\hat{\sigma}^2_{\tau\beta}$ are small; in fact, they are not significantly different from zero.

The Mixed Model

Now suppose that one of the factors, *A*, is fixed and the other, *B*, is random. This is called the **mixed model** analysis of variance. The linear model is

$$Y_{ijk} = \mu + \tau_i + \beta_j + (\tau\beta)_{ij} + \epsilon_{ijk} \begin{cases} i = 1, 2, \ldots, a \\ j = 1, 2, \ldots, b \\ k = 1, 2, \ldots, n \end{cases} \tag{VII-7}$$

In this model, τ_i is a fixed effect defined such that $\sum_{i=1}^{a} \tau_i = 0$, β_j is a random effect, the interaction term $(\tau\beta)_{ij}$ is a random effect, and ϵ_{ijk} is a normally and independently distributed random variable with mean zero and variance σ^2. The interaction elements $(\tau\beta)_{ij}$ are normal random variables with mean zero and variance $[(a - 1)/a]\sigma^2_{\tau\beta}$, and $\sum_{i=1}^{a} (\tau\beta)_{ij} = 0$. Because the sum of the interaction effects over the levels of the fixed factor equals zero, this version of the mixed model is called the **restricted model.** The interaction elements are not all independent. For more details, see Montgomery (1997).

The expected mean squares in this case are

$$\begin{aligned} E(MS_A) &= \sigma^2 + n\sigma^2_{\tau\beta} + \frac{bn\sum_{i=1}^{a}\tau_i^2}{a-1} \\ E(MS_B) &= \sigma^2 + an\sigma^2_{\beta} \\ E(MS_{AB}) &= \sigma^2 + n\sigma^2_{\tau\beta} \\ E(MS_E) &= \sigma^2 \end{aligned} \tag{VII-8}$$

Therefore, the appropriate test statistic for testing H_0: $\tau_i = 0$ is

$$F_0 = \frac{MS_A}{MS_{AB}} \tag{VII-9}$$

for which the reference distribution is $F_{a-1,(a-1)(b-1)}$. For testing H_0: $\sigma^2_\beta = 0$, the test statistic is

$$F_0 = \frac{MS_B}{MS_E} \tag{VII-10}$$

for which the reference distribution is $F_{b-1,ab(n-1)}$. Finally, for testing H_0: $\sigma^2_{\tau\beta} = 0$ we would use

$$F_0 = \frac{MS_{AB}}{MS_E} \tag{VII-11}$$

for which the reference distribution is $F_{(a-1)(b-1),ab(n-1)}$.

The variance components σ^2_β, $\sigma^2_{\tau\beta}$, and σ^2 may be estimated by eliminating the first equation from equations VII-8, leaving three equations in three unknowns, whose solution is

$$\hat{\sigma}^2_\beta = \frac{MS_B - MS_E}{an}$$

$$\hat{\sigma}^2_{\tau\beta} = \frac{MS_{AB} - MS_E}{n}$$

and

$$\hat{\sigma}^2 = MS_E$$

This general approach can be used to estimate the variance components in *any* mixed model. After eliminating the mean squares containing fixed factors, there will always be

Table VII-3 Analysis of Variance for the Two-Factor Mixed Model (A fixed, *B* random)

Source of Variation	Sum of Squares	Degrees of Freedom	Expected Mean Square	Mean Square	F_0
Rows (A)	SS_A	$a-1$	$\sigma^2 + n\sigma_{\tau\beta}^2 + bn\sum\tau_i^2/(a-1)$	MS_A	MS_A/MS_{AB}
Columns (B)	SS_B	$b-1$	$\sigma^2 + an\sigma_\beta^2$	MS_B	MS_B/MS_E
Interaction	SS_{AB}	$(a-1)(b-1)$	$\sigma^2 + n\sigma_{\tau\beta}^2$	MS_{AB}	MS_{AB}/MS_E
Error	SS_E	$ab(n-1)$	σ^2	MS_E	
Total	SS_T	$abn-1$			

a set of equations remaining that can be solved for the variance components. Table VII-3 summarizes the analysis of variance for the two-factor mixed model.

Computer Output

Some software packages have the capability to handle random factors in the analysis of variance (SAS, Minitab, Statgraphics, for example). Table VII-4 is the computer solution to Example VII-1 from Minitab. Notice that the *F*-tests on the main effects have been constructed using MS_{AB} in the denominator. The results closely match those in Table VII-2.

Table VII-5 is the Minitab output for the breaking strength data assuming a mixed model, where operators are fixed and machines are random. Minitab assumes that the restricted form of the mixed model has been specified [not all software packages make this assumption—SAS uses an unrestricted model, that is, the term $\sum_{i=1}^{a} (\tau\beta)_{ij} \neq 0$ in general]. Notice that the test statistics for main effects and interactions have been constructed properly using equations VII-9 through VII-11.

Table VII-4 Minitab Output for Example VII-1

Analysis of Variance (Balanced Designs)

Factor	Type	Levels	Values			
Operator	random	3	A	B	C	
Machine	random	4	1	2	3	4

Analysis of Variance for Strength

Source	DF	SS	MS	F	P
Operator	2	26.333	13.167	2.21	0.190
Machine	3	24.333	8.111	1.36	0.340
Operator*Machine	6	35.667	5.944	1.59	0.234
Error	12	45.000	3.750		
Total	23	131.333			

Table VII-5 Minitab Analysis of Variance for the Breaking Stength Data in Table VII-1 Where Operators Are Fixed and Machines Are Random (Mixed Model)

Analysis of Variance (Balanced Designs)

Factor	Type	Levels	Values			
Operator	fixed	3	A	B	C	
Machine	random	4	1	2	3	4

Analysis of Variance for Strength

Source	DF	SS	MS	F	P
Operator	2	26.333	13.167	2.21	0.190
Machine	3	24.333	8.111	2.16	0.145
Operator*Machine	6	35.667	5.944	1.59	0.234
Error	12	45.000	3.750		
Total	23	131.333			

EXERCISES

VII-1. Consider the experiment described in Example 13-4. Suppose that both factors were random.
(a) Analyze the data and draw appropriate conclusions.
(b) Estimate the variance components.

VII-2. For the breaking strength data in Table VII-1, suppose that the operators were chosen at random, but machines were a fixed factor. Does this influence the analysis or your conclusions?

VII-3. A company employs two time-study engineers. Their supervisor wishes to determine whether the standards set by them are influenced by an interaction between engineers and operators. She selects three operators at random and conducts an experiment in which the engineers set standard times for the same job. She obtains the data shown here.

	Operator		
Engineer	1	2	3
1	2.59	2.38	2.40
	2.78	2.49	2.72
2	2.15	2.85	2.66
	2.86	2.72	2.87

(a) State the appropriate hypotheses.
(b) Use the analysis of variance to test these hypotheses with $\alpha = 0.05$.
(c) Graphically analyze the residuals from this experiment.
(d) Estimate the appropriate variance components.

VII-4. Consider the experiment on baked anode density described in Exercise 13-4. Suppose that positions on the furnace were chosen at random and temperature is a fixed factor.
(a) State the appropriate hypotheses.
(b) Use the analysis of variance to test these hypotheses with $\alpha = 0.05$.
(c) Estimate the variance components.

VII-5. Consider the experiment described in Exercise VII-3. How does the analysis (and conclusions) change if both factors are random? Use $\alpha = 0.05$.

Appendix C
Bibliography

INTRODUCTORY WORKS AND GRAPHICAL METHODS

Chambers, J., Cleveland, W., Kleiner, B., and P. Tukey (1983), *Graphical Methods for Data Analysis,* Wadsworth & Brooks/Cole, Pacific Grove, CA. A very well-written presentation of graphical methods in statistics.

Freedman, D., Pisani, R., Purves R., and A. Adbikari (1991), *Statistics,* 2nd edition, Norton, New York. An excellent introduction to statistical thinking, requiring minimal mathematical background.

Hoaglin, D., Mosteller, F., and J. Tukey (1983), *Understanding Robust and Exploratory Data Analysis,* John Wiley & Sons, New York. Good discussion and illustration of techniques such as stem-and-leaf displays and box plots.

Tanur, J., et al. (eds.) (1989), *Statistics: A Guide to the Unknown,* 3rd edition, Wadsworth & Brooks/Cole, Pacific Grove, CA. Contains a collection of short nonmathematical articles describing different applications of statistics.

Tukey, J. (1977), *Exploratory Data Analysis,* Addison-Wesley, Reading, MA. Introduces many new descriptive and analytical methods. Not extremely easy to read.

PROBABILITY

Derman, C., Gleser, L., and I. Olkin (1980), *Probability Models and Applications,* Macmillan, New York. A comprehensive treatment of probability at a higher mathematical level than this book.

Hoel, P. G., Port, S. C., and C. J. Stone (1971), *Introduction to Probability Theory,* Houghton Mifflin, Boston. A well-written and comprehensive treatment of probability theory and the standard discrete and continuous distributions.

Mosteller, F., Rourke, R., and G. Thomas (1970), *Probability with Statistical Applications,* 2nd ed., Addison-Wesley, Reading, MA. A precalculus introduction to probability with many excellent examples.

Ross, S. (1985), *A First Course in Probability,* 3rd ed., Macmillan, New York. More mathematically sophisticated than this book, but has many excellent examples and exercises.

MATHEMATICAL STATISTICS

Efron, B., and R. Tibshirani (1993), *An Introduction to the Bootstrap,* Chapman and Hall, New York. An important reference on this useful but computer-intensive technique.

Hoel, P. G. (1984), *Introduction to Mathematical Statistics,* 5th ed., John Wiley & Sons, New York. An outstanding introductory book, well written, and generally easy to understand.

Hogg, R., and A. Craig (1978), *Introduction to Mathematical Statistics,* 4th ed., Macmillan, New York. Another classic work on the mathematical principles of statistics; higher level than the Hoel book, but contains excellent discussions of estimation and hypothesis testing.

Larsen, R., and M. Marx (1986), *Introduction to Mathematical Statistics,* 2nd ed., Prentice-Hall, Englewood Cliffs, NJ. Written at a relatively low mathematical level, very readable.

Larson, H. J. (1982), *Introduction to Probability Theory and Statistical Inference,* 3rd ed., John Wiley & Sons, New York. An extremely well-written book that gives broad coverage to many aspects of probability and mathematical statistics.

ENGINEERING STATISTICS

Devore, J. L. (1995), *Probability and Statistics for Engineering and the Sciences,* 4th ed., Wadsworth & Brooks/Cole, Pacific Grove, CA. Covers many of the same topics as this text, but at a slightly higher mathematical level. Many of the examples and exercises involve applications to biological and life sciences.

Hines, W. W., and D. C. Montgomery (1990), *Probability and Statistics in Engineering and Management Science,* 3rd ed., John Wiley & Sons, New York. Covers many of the same topics as this book. More emphasis on probability and a higher mathematical level.

Ross, S. (1987), *Introduction to Probability and Statistics for Engineers and Scientists,* John Wiley & Sons, New York. More tightly written and mathematically oriented than this book, but contains some good examples.

Walpole, R. E., Myers, R. H., and S. L. Myers (1998), *Probability and Statistics for Engineers and Scientists,* 6th ed., Prentice-Hall, Inc., Upper Saddle River, New Jersey. A very well-written book at about the same level as this one.

REGRESSION ANALYSIS

Daniel, C., and F. Wood (1980), *Fitting Equations to Data,* 2nd ed., John Wiley & Sons, New York. An excellent reference containing many insights on data analysis.

Draper, N., and H. Smith (1998), *Applied Regression Analysis,* 3rd ed., John Wiley & Sons, New York. A comprehensive book on regression written for statistically oriented readers.

Montgomery, D. C., and E. A. Peck (1992), *Introduction to Linear Regression Analysis,* 2nd ed., John Wiley & Sons, New York. A comprehensive book on regression written for engineers and physical scientists.

Myers, R. H. (1990), *Classical and Modern Regression with Applications,* 2nd ed., PWS-Kent, Boston. Contains many examples with annotated SAS output. Very well written.

Neter, J., Wasserman, W., Nachtsheim, C., and M. Kutner (1996), *Applied Linear Statistical Models,* 4th ed., Richard D. Irwin, Homewood, Ill. The first part of the book is an introduction to simple and multiple linear regression. The orientation is to business and economics.

Younger, M. S. (1985), *A Handbook for Linear Regression,* 2nd ed., Duxburg, Boston. A good presentation of regression methods. The discussion of SAS, BMD, and SPSS computer packages is excellent.

DESIGN OF EXPERIMENTS

Box, G. E. P., Hunter, W. G., and J. S. Hunter (1978), *Statistics for Experimenters,* John Wiley & Sons, New York. An excellent introduction to the subject for those readers desiring a statistically oriented treatment. Contains many useful suggestions for data analysis.

Mason, R. L., Gunst, R. F., and J. F. Hess (1989), *Statistical Design and Analysis of Experiments,* John Wiley & Sons, New York. A comprehensive book covering basic statistics, hypothesis testing and confidence intervals, elementary aspects of experimental design, and regression analysis.

Montgomery, D. C. (1997), *Design and Analysis of Experiments,* 4th ed., John Wiley & Sons, New York. Written at the same level as the Box, Hunter, and Hunter book, but focused on engineering applications.

NONPARAMETRIC STATISTICS

Conover, W. J. (1980), *Practical Nonparametric Statistics,* 2nd ed., John Wiley & Sons, New York. An excellent exposition of the methods of nonparametric statistics, many good examples and exercises.

Hollander, M., and D. Wolfe (1973), *Nonparametric Statistical Methods,* John Wiley & Sons, New York. A good reference book, with a very useful set of tables.

STATISTICAL QUALITY CONTROL AND RELATED METHODS

Duncan, A. J. (1986), *Quality Control and Industrial Statistics,* 5th ed., Richard D. Irwin, Homewood, Illinois. A classic book on the subject.

Grant, E. L., and R. S. Leavenworth (1988), *Statistical Quality Control,* 6th ed., McGraw-Hill, New York. One of the first books on the subject; contains many good examples.

John, P. W. M. (1990), *Statistical Methods in Engineering and Quality Improvement,* John Wiley & Sons, New York. Not a methods book, but a well-written presentation of statistical methodology for quality improvement.

Montgomery, D. C. (1996), *Introduction to Statistical Quality Control,* 3rd ed., John Wiley & Sons, New York. A modern comprehensive treatment of the subject written at the same level as this book.

Nelson, W. (1982), *Applied Life Data Analysis,* John Wiley & Sons, New York. Contains many examples of using statistical methods for the study of failure data; a good reference for the statistical aspects of reliability engineering and the special probability distributions used in that field.

Ryan, T. P. (1989), *Statistical Methods for Quality Improvement,* John Wiley & Sons, New York. Gives broad coverage of the field, with some emphasis on newer techniques.

Wadsworth, H. M., Stephens, K. S., and A. B. Godfrey (1986), *Modern Methods for Quality Control and Improvement,* John Wiley & Sons, New York. A comprehensive treatment of statistical methods for quality improvement at a somewhat higher level than this book.

Western Electric Company (1956), *Statistical Quality Control Handbook,* Western Electric Company, Inc., Indianapolis, Indiana. An oldie but a goodie.

Appendix D
Answers to Selected Exercises

CHAPTER 10

Section 10-2

10-1. **a.** $\hat{\beta}_0 = 48.013$, $\hat{\beta}_1 = -2.33$
b. 37.99 **c.** 39.39 **d.** 6.71

10-3. **a.** $\hat{\beta}_0 = 0.4631476$, $\hat{\beta}_1 = 0.0074902$
b. $\hat{\beta}_1 = 0.00749$

10-5. **a.** $\hat{\beta}_0 = 13.3202$, $\hat{\beta}_1 = 3.32437$
b. 38.253 **c.** -2.02732

10-7. **a.** $\hat{\beta}_0 = 33.5348$, $\hat{\beta}_1 = 0.0353971$
b. 28.26 **c.** 1.50048

10-9. **b.** $\hat{\beta}_0 = -9.8131$, $\hat{\beta}_1 = 0.171484$
c. 4.76301

10-11. **b.** $\hat{\beta}_0 = 0.470467$, $\hat{\beta}_1 = 20.5673$
c. 21.038 **d.** 1.6629

10-17. $\hat{\beta}_0 = 0$, $\hat{\beta}_1 = 21.031461$

Section 10-5

10-19. **a.** $f_0 = 73.95$, reject H_0 **b.** 0.00796
c. $se(\hat{\beta}_0) = 0.04095$,
$se(\hat{\beta}_1) = 0.000484$

10-21. **a.** $t_0 = 8.518$, reject H_0
b. $f_0 = 72.5563$, reject H_0
c. 8.76775, $se(\hat{\beta}_0) = 2.57172$,
$se(\hat{\beta}_1) = 0.390276$
d. $t_0 = 5.1795$, reject H_0

10-23. **a.** $f_0 = 4.53$, do not reject H_0
b. 3.6596, $se(\hat{\beta}_0) = 2.614$,
$se(\hat{\beta}_1) = 0.0166281$
c. $t_0 = 0.87803$, do not reject H_0
d. $t_0 = 12.83$, reject H_0

10-25. **a.** $f_0 = 44.6567$, reject H_0,
P-value $= 0.000003$
b. 1.98
c. $se(\hat{\beta}_0) = 2.1356$,
$se(\hat{\beta}_1) = 0.0256613$
d. $t_0 = -4.5957$, reject H_0,
P-value $= 0.00022$

10-27. **a.** $f_0 = 155$, reject H_0
b. 9811.21, $se(\hat{\beta}_0) = 2.9668$,
$se(\hat{\beta}_1) = 45.3468$
c. $t_0 = -2.3466$, do not reject H_0,
P-value $= 0.0306$
d. $t_0 = 57.8957$, reject H_0
e. $t_0 = 2.765$, do not reject H_0

Section 10-6 and 10-7

10-31. **a.** $(-2.917478, -1.7421254)$
b. (46.714, 49.311)
c. (41.284, 43.093)
d. (39.095, 45.2826)

10-33. **a.** $(-0.00961, -0.00444)$
b. (16.2448, 27.3318)
c. (7.914, 10.3717)
d. (4.072, 14.214)

10-35. **a.** (9.101, 9.315)
b. $(-11.622, -1.049)$
c. (498.72, 501.53)
d. (495.57, 504.67)

10-37. **a.** (0.03689, 0.10183)
b. (44.0897, 49.1185)
c. (44.0897, 49.1185)
d. (37.8298, 55.3784)

10-39. **a.** (201.552, 266.59)
b. $(-4.67, -2.347)$
c. (111.834, 145.79)

10-41. **a.** $(-43.1964, -30.7272)$
b. (2530.09, 2720.68)
c. (1823.783, 1948.525)
d. (9.8201, 32.2559)

10-43. **d.** $R^2 \equiv 76.73\%$

10-45. **a.** $R^2 = 82.9\%$ **c.** Yes

10-47. **a.** $R^2 = 71.27\%$
b. $f_0 = 0.196$, do not reject H_0

10-49. **a.** $R^2 = 85.22\%$ **b.** $SS_{PE} = 4.505$
c. $f_0 = 8.864$, do not reject H_0

Section 10-10

10-55. **a.** $\hat{\beta}_0 = -0.0280411$, $\hat{\beta}_1 = 0.990987$
b. $f_0 = 4.41$, reject H_0 **c.** 0.903
d. $t_0 = 8.9345$, reject H_0
e. $z_0 = 3.879$, reject H_0
f. (0.7677, 0.9615)

10.57. **a.** -0.738
b. Reject H_0, P-value $= 0.00000738$
c. $(-0.871, -0.504)$
d. Do not reject H_0, P-value $= 0.6936$

10-59. **a.** Reject H_0, P-value $= 0$
b. (0.3358, 0.8007) **c.** Yes

10-61. **a.** 0.933 **b.** $t_0 = 10.06$, reject H_0
c. $\hat{\beta}_0 = 0.72538$, $\hat{\beta}_1 = 0.498081$,
$f_0 = 101.16$, reject H_0
d. $t_0 = 0.468345$, do not reject H_0

Supplemental Exercises

10-65. **a.** $\hat{\beta}_0 = 93.34$, $\hat{\beta}_1 = 15.64$
b. $f_0 = 4.60$, reject H_0
c. (7.961, 23.322)
d. (74.758, 111.923)
e. (26.18, 138.70)
f. $f_0 = 2.002$, do not reject H_0

10-67. **b.** $\hat{\beta}_0 = -0.8819$, $\hat{\beta}_1 = 0.00385$
c. $f_0 = 122.03$, reject H_0 **d.** No
e. $\hat{\beta}_0 = 0.5967$, $\hat{\beta}_1 = 0.00097$, yes

10-69. $\hat{\beta}_1 = 0.7916$

CHAPTER 11

Sections 11-2 and 11-3

11-1. **b.** $\hat{\beta} = \begin{pmatrix} 171.054 \\ 3.713 \\ -1.126 \end{pmatrix}$ **c.** 189.48

11-5. **a.** $\hat{y} = 33.4491 - 0.05435x_1 + 1.07822x_2$
b. 21.457 mpg

11-7. **a.** $\hat{y} = 350.9943 - 1.272x_1 - 0.1539x_2$
b. 165.29
c. $\hat{y}' = 125.8655 - 7.7586x_1 - 0.0943x_2 - 0.0092x_1x_2$
d. 184.13

11-9. **a.** $\hat{y} = 47.174 - 9.7352x_1 + 0.4283x_2 + 18.2375x_3$
b. 91.43

11-11. $\hat{y} = -8.0119 + 0.494x_1 + 0.0018x_2 + 0.0023x_3 + 0.0383x_4 - 0.2868x_5 - 0.0128x_6 + 0.030x_7 + 0.0407x_8 - 0.2083x_9$

Sections 11-4 and 11-5

11-13. **a.** $f_0 = 260$, reject H_0 **b.** 8.517
c. $\text{se}(\hat{\beta}_2) = 0.0861$, $\text{se}(\hat{\beta}_1) = 0.195$
d. Reject H_0, both significant

11-15. **a.** $f_0 = 25.7465$, reject H_0, P-value < 0.000001
b. 3.23 **c.** 0.000714
d. Reject H_0, all variables significant

11-17. **a.** $f_0 = 53.316$, reject H_0
b. Only x_1 is significant

11-19. **a.** $f_0 = 5.106$, do not reject H_0
b. 242.716 **c.** Do not reject H_0

11-21. **a.** $f_0 = 7.714$, do not reject H_0
b. $f_0 = 1.47$, do not reject H_0
c. 561.28, 650.14

11-23. **a.** $f_0 = 840.55$, reject H_0
b. 12.1078 **c.** x_2 is not significant

11-25. **a.** $f_0 = 101.79$, reject H_0
b. PTS is significant

Sections 11-6 and 11-7

11-27. **a.** $(-0.00657, -0.00122)$
b. 0.49765 **c.** (7.16, 9.22)

11-29. **a.** $-0.26718 \le \beta_1 \le 1.47793$;
$-3.61373 \le \beta_2 \le 21.4610$;
$-4.21947 \le \beta_3 \le 7.09438$;
$-1.72211 \le \beta_4 \le 1.74932$
b. (263.78, 311.34)
c. (243.711, 331.413)

11-31. **a.** β_2: $(-0.595, 0.535)$;
β_3: (0.229, 0.812);
β_4: $(-0.216, 0.013)$;
β_5: $(-7.298, 2.977)$
b. (7.982, 10.009)
c. (6.8481, 11.143)

11-33. **a.** Temp: $(-0.00003, 0.00012)$;
Soaktime: (0.00203, 0.00288);
Soakpct: $(-0.02306, 0.05976)$;
Dftime: (0.00501, 0.01056);
Diffpct: $(-0.01969, 0.01342)$
b. (0.0206, 0.0234)

11-35. **a.** (0.3882, 0.5998)
b. $\hat{y}' = -5.767703 + 0.496501x_{\text{pts}}$
c. (0.4648, 0.5282)

Section 11-8

11-37. **a.** 0.82897

11-39. **a.** 0.86179 **b.** 0.920453

11-41. **b.** 99.37%
c. $\hat{y}' = 6.22489 - 0.16647x_1 - 0.000228x_2 + 0.157312x_3$

11-43. **a.** 95.82% **b.** Normality is acceptable
c. Last observation is influential

11-45. **a.** 0.12 **b.** Points 17 and 18

Sections 11-9 and 11-10

11-47. **a.** $\hat{y}' = -1.633 + 1.232x - 1.495x^2$
b. $f_0 = 1{,}858{,}613$, reject H_0
c. $t_0 = -601.64$, reject H_0

11-49. **a.** 802.943
b. $\hat{y}' = -26{,}204 + 189.09x - 0.331x^2$
c. They are the same

11-51. **a.** $\hat{y}' = -1.769 + 0.421x_1 + 0.222x_2 - 0.128x_3 - 0.02x_1x_2 + 0.009x_1x_3 + 0.003x_2x_3 - 0.019x_1^2 - 0.007x_2^2 + 0.001x_3^2$
b. Reject H_0
d. $f_0 = 1.612$, do not reject H_0

Section 11-11

11-55. **a.** Min. $MS_E = 6.58$, $C_p = 5.88$: $x_2, x_3, x_4, x_5, x_7, x_8, x_{10}$
Min $C_p = 5.02$, $MS_E = 7.97$: x_5, x_8, x_{10}
b. $\hat{y}' = 34.434 - 0.048x_1$, $MS_E = 8.81$, $C_p = 5.55$
c. Same as part b
d. $\hat{y}' = 0.341 + 2.862x_5 + 0.246x_8 - 0.010x_{10}$, $MS_E = 7.97$, $C_p = 5.02$

11-57. **a.** $\hat{y} = 4.656 + 0.511x_3 - 0.124x_4$
b. Same as part a **c.** Same as part a
d. All models are the same

11-59. **a.** $\hat{y} = -0.304 + 0.083x_1 - 0.031x_3 + 0.0004x_2^2$, $C_p = 4.04$, $MS_E = 0.004$
b. $\hat{y} = -0.256 + 0.783x_1 + 0.022x_2 - 0.042x_3 + 0.0008x_3^2$, $C_p = 4.66$, $MS_E = 0.004$

11-61. **a.** $\hat{y} = -3.517 + 0.486x_1 - 0.156x_9$, $C_p = -1.67$
b. $\hat{y} = -5.964 + 0.495x_1 + 0.025x_7 - 0.163x_9$, $MS_E = 1.67$
c. $R^2 = 0.98448$, yes

11-63. **a.** Yes, $C_p > p$
b. Yes, both MS_E and C_p are reduced

Supplemental Exercises

11-65. **a.** $f_0 = 1323.62$, reject H_0
b. x_4 is significant

11-67. **a.** $\hat{y}' = -0.908 + 5.482x_1^* + 1.126x_2^* - 3.920x_3^* - 1.143x_4^*$
b. x_1, x_2, x_3, x_4 all significant

11-71. Yes, VIF's for x_3 and x_5 exceed 10

11-73. **a.** Reject H_0
b. Do not reject H_0 **c.** No

CHAPTER 12

Section 12-2

12-1. **a.** Reject H_0
c. Model is satisfactory

12-3. **a.** Reject H_0 **b.** 0.0005

12-5. **a.** Reject H_0
c. (141.96, 148.05), (7.36, 24.14)

12-7. **a.** Do not reject H_0 **b.** 0.2138

12-9. **a.** Reject H_0 **b.** 0.0022
d. (71.02, 79.98) **e.** (8.423, 26.327)

Section 12-3

12-11. 3.32 3.93 4.42

12-13. 2666.25 2933.75 2971.00 3156.25

12-15. 17.4 20.8 28.2

12-17. 75.5 82.125 92.875

12-23. 129.25 131.5 145 145.25 145.25

12-29. 12.3 13.38 14.8 14.8

Section 12-4

12-31. **a.** Reject H_0 **b.** 0.0141 **c.** 0.0148
d. Residuals are acceptable

12-33. **a.** Do not reject H_0 **b.** 0
c. 23.9994
d. Lower variablity in Chemical 4

Section 12-5

12-35. **a.** Reject H_0
c. Some indication of greater variability at the lowest nozzle

12-37. **a.** Do not reject H_0
b. Some indication of variability increasing magnitude of the response

12-39. **a.** Do not reject H_0
b. Observation 20 is unusual

Section 12-6

12-41. $n = 5$

12-43. 0.6

12-45. $n = 5$

Supplemental Exercises

12-47. **a.** Reject H_0 **b.** 0.56 is an outlier

12-49. **a.** Do not reject H_0 **b.** 0.2107
d. 4.037, 0.342

12-51. **a.** Reject H_0
b. -0.19 $\underline{1.14\ 1.34}$ $\underline{1.74\ 1.82\ 2.12}$

12-53. **a.** 0.3 **b.** $n \cong 11$

CHAPTER 13

Section 13-4

13-1. **a.**

Factor	Temp	Mat	Temp × Mat
f_0	28.968	7.911	3.56

c. Residuals are satisfactory

13-3. **a.**

Factor	Glass	Phos.	Glass × Phos
f_0	273.8	8.842	1.263

13-9. **a.**

Factor	Doping	Anneal	Doping × Anneal
f_0	25.231	543.52	3.54

Section 13-5

13-11. **a.** A—Hardwood B—Cooking Time
C—Freeness

Factor	A	B	C	AB	AC	BC	ABC
f_0	7.553	31.313	19.706	2.889	2.937	0.947	0.940
P-value	0.0042	0	0	0.0816	0.0495	0.4063	0.4634

Section 13-6

13-13. **a.** Variables B, C, and AC are significant
b. $\hat{y} = 413.125 + 9.124x_A + 42.125x_B + 35.875x_C - 59.625_{AC}$

13-17. **b.** The important affects are A, B, D, E, AB.
c. A, B, D, E should be increased.

13-19. **a.** The important variables are A, B, and AB.
d. Residuals from the model using A, B, and AB are acceptable.

13-21. **a.** $F = 84.64$, P-value $= 0.0008$, reject H_0
b. 8.20

Section 13-7

13-23. **a.** Block 1: (1) ab ac bc;
Block 2: a b c abc;
A = 13.75, B = 127.75,
C = 97,75, AB = −21.25,
AC = −137.25, BC = −52.25

13-25. Block 1: (1) abc ad bc;
Block 2: ac abd cd;
Block 3: ab c bd acd;
Block 4: a bc d abcd

13-27. ABC, CDE, and ABDE

13-29. **a.** Significant variables are C, AB, AC
b. Residuals from model using A, B, C, AB, and AC are acceptable.
c. The second replicate could have confounded a different effect with blocks.
d. Confound ABC in Replicable 1 and a two-factor interaction in Replicate 2.

Section 13-8

13-31. **a.** Important effects are:
D + ABCE = 4.545;
A + BCDE = 1.435;
B + ACDE = −1.465;
AB + CDE = 1.15;
AD + BCD
b. Residuals from model using A, B, D, AB, and AD are acceptable.
c. Collapsing on C and E results in two replicates of a full factorial in A, B, and D.

13-33. **c.** A: −1.525, B: −5.175, C: 2.275, D: −0.675, E: 2.275
d. AB and AD are available to use as error since they are found insignificant.

13-35. Collapsed design is two replicates of a 2^{4-1} design because ABCE is a word in the complete defining relation.

13-37. A + BD + CE + ABCDE = 1450;
B + AD + CDE + ABCE = 3650;
C + AE + BDE + ABCD = −150;
D + AB + BCE + ACDE = 1750;
E + AC + BCD + ABDE = 650;
BC + DE + ABE + ACD = 250;
BE + CD + ABC + ADE = −50

13-39. A + BD + CE; B + AD + CF;
C + AE + BF; D + AB + EF;
E + AC + DF; F + BC + DE;
AF + BE + CD

Section 13-9

13-41. **b.** $\hat{y} = 79.947 - 1.115x_1 - 4.208x_2 - 1.801x_1x_2 + 5.707x_1^2 - 5.743x_2^2$, $f_0 = 1.67$, run number 6 is an outlier.
c. The yield increases as x_1 is decreased and x_2 is near its zero level.

13-43. **a.** $1 - 2.5x_1 < x_2 < 2.25 - 1.5x_1$
b. Let $x_1 = 1.5$, $x_2 = -1.5$

13-45. **a.** $\hat{y} = 327.57 + 177x_A + 109.425x_B + 131.44x_C + 66x_{AB} + 75.47x_{AC} + 43.58x_{BC}$
b. $\hat{y} = 48 + 11.53x_A + 15.32x_B + 29.19x_C$

Supplemental Exercises

13-47. **a.**

Factor	Gear	Time	Gear×Time
f_0	3.905	25.657	0.260

c. Residuals are acceptable

13-49. **a.**

Factor	Bump	Align	Bump×Align
f_0	3.89	10.06	0.875

b. Bump at 75 or 130 with method 3—to minimize the response.

13-51. **b.** Important effects are A = 5.7625, B = 5.5625, C = 6.2375
c. $MS_E = 14.03$
d. $\hat{y} = 72.89 + 2.88x_A + 2.78x_B + 3.19x_C$
e. Variability is greater when A is at its high level

13-55. Generators: E = BCD; F = ACD; G = ABC; H = ABD

13-57. **a.** Important effects are A, C, D, AC, and AD
b. $MS_E = 1.625$
c. Residuals are acceptable

13-59. **a.** D = ABC, main effects are aliased with three-factor interactions except E is aliased with ABCDE. Other aliases are Ab + Cd, AC + BD, AD + BC when higher-order effects are ignored.

CHAPTER 14

Section 14-2

14-1. No, P-value = 0.109

14-3. Yes, P-value = 0.0002

14-5. **a.** P-value = 0.151
b. P-value = 0.099 without continuity correction, 0.151 with continuity correction.

14-7. Without continuity correction: P-value = 0.250, with continuity correction: 0.359

14-9. P-value = 0.289, do not reject H_0

14-11. Without continuity correction: P-value = 0.00047, with continuity correction: 0.0019

14-13. Reject H_0

14-15. Without continuity correction: P-value = 0.0004, with continuity correction: 0.0019

14-17. **a.** 0.0025 **b.** 0.115
c. The sign test that rejects if $R^- \leq 1$ has an $\alpha = 0.011$ based on the binomial distribution.
d. 0.487

14-19. P-value = 0.0075

Section 14-3

14-21. $w = 80.5$, do not reject H_0

14-23. $w = 5$, reject H_0

14-25. $w = 17.5$, reject H_0

14-27. $w = 1$, reject H_0

Section 14-4

14-29. $w = 38$, do not reject H_0

14-31. P-value = 0.716

14-33. $w = 73$, reject H_0

14-35. P-value = 0.13

Section 14-5

14-37. Reject H_0

14-39. Do not reject H_0

14-41. P-value = 0.018

Supplementary Exercises

14-43. $w = 5$, do not reject H_0

14-45. P-value = 0.133, do not reject H_0

14-47. $r = 1$, do not reject H_0

14-49. $w = 4$, reject H_0

14-51. $w = 55$, reject H_0

14-53. P-value = 0, reject H_0

14-55. Do not reject H_0

14-57. Reject H_0

CHAPTER 15

Section 15-5

15-1. **a.** $\bar{x}$ Chart: UCL = 37.5789, CL = 34.32, LCL = 31.0611, R chart UCL = 11.9461, CL = 5.65, LCL = 0
b. One point beyond the control limits on the $\bar{x}$ Chart Revised control limits:
$\bar{x}$ Chart: UCL = 37.4038, CL = 34.0947, LCL = 30.7857, R chart UCL = 12.1297, CL = 5.73684, LCL = 0

15-3. **a.** $\bar{x}$ Chart: UCL = 15.6299, CL = 10.7125, LCL = 5.79506, R = chart UCL = 15.4047, CL = 6.75, LCL = 0
b. No points are beyong the control limits

15-5. **a.** $\bar{x}$ Chart: UCL = 242.78, CL = 223, LCL = 203.22, R chart UCL = 72.51, CL = 34.286, LCL = 0
b. Mean: 223, standard deviation: 14.74

15-7. **a.** $\bar{x}$ Chart: UCL = 7.44253, CL = 6.2836, LCL = 5.12467, R chart UCL = 2.9153, CL = 1.1328, LCL = 0
b. No points are beyond the control limits.

Section 15-6

15-9. **a.** x Chart: UCL = 92.1893, CL = 81.6667, LCL = 71.144 MR Chart: UCL = 12.9324, CL = 3.95652, LCL = 0 the process is in control.
b. Mean: 81.6667, standard deviation: 3.508

15-11. **a.** x Chart: UCL = 552.12, CL = 500.6, LCL = 449.088 MR Chart: UCL = 63.308, CL = 19.3684, LCL = 0 The process is in control.
b. Mean: 500.6, standard deviation: 17.17

Section 15-7

15-13. **a.** PCR = 1.13, PCR_K = 1.10 **b.** 0.075%

15-15. PCR = 0.508, process capability is poor

15-17. PCR = 0.90, PCR_K = 0.84, the process capability is marginal

15-19. PCR = 0.50, PCR_K = 0.36, process capability is poor

15-21. PCR = 0.49, PCR_K = 0.47, process capability is poor

Section 15-8

15-23. **a.** U Chart: UCL = 3.81116, CL = 1.94167, LCL = 0.0721748
b. There are two points out of control. The revised control limits are UCL = 3.463, CL = 1.709, LCL = 0

15-25. **a.** C Chart: UCL = 19.0558, CL = 9.708, LCL = 0.360874 There are two points out of control just as with the U chart.

Section 15-9

15-27. **a.** 0.218 **b.** 4.59

15-29. **a.** 0.2946 **b.** 1.42

15-31. **a.** 0.2451 **b.** 4.08

15-33. **a.** 0.96327 **b.** 1.04

15-35. **a.** 0.16603 **b.** 6.02

Section 15-10

15-37. **a.** 0.169548
b. H = 0.85, K = 0.0848

15-39. **a.** 38 **b.** 10.4

Supplemental Exercises

15-41. **a.** $\bar{x}$ Chart: UCL = 64.0179, CL = 64.0002, LCL = 63.9862 R chart UCL = 0.0443678, CL = 0.01724, LCL = 0
b. Means: 64.0002, standard deviation: 0.0102
c. PCR = 1.63 **d.** PCR_K = 1.63
e. 0.025 **f.** 12.18

15-43. **a.** UCL = 4.81991, CL = 2.64, LCL = 0.460092
b. Process is out of control. The revised control limits are UCL = 4.54206, CL = 2.44444, LCL = 0.346827
c. UCL = 4.18143, CL = 2.64, LCL = 1.09857
Process is out of control. The revised control limits are UCL = 3.51897, CL = 2.13333, LCL = 0.747693

15-45. **a.** 0.125 **b.** 0.004

15-49. **a.** 0.10 **b.** 0.09 **c.** 0.2133

15-51. **a.** 0.085 **b.** 0.444

15-55. 0.00135

15-57. $K = 8.5853$, $H = 85.85$

Appendix B-I

I-1. 120

I-3. 144

I-5. 14,400

I-7. **a.** 416965528 **b.** 113588800

I-9. **a.** 21 **b.** 2520 **c.** 720

I-11. **a.** 1000 **b.** 160 **c.** 720

I-13. 0.461

I-15. **a.** 0.0082 **b.** 0.0082

Appendix B-II

II-1. **b.** $(m + 1)/2$, $(9m^2 + 10m - 13)/12$

II-3. **b.** $(1 - p)/p^2$

II-5. $E(X) = 1/4$ $V(X) = 5/16$

II-7. **b.** $E(X) = 2/\lambda^2$, $V(X) = 1/\lambda^2$

II-9. **a.** $(1 - t/\lambda)^{-n}$ **b.** Gamma

Appendix B-III

III-1. 1/4 at $y = 3, 5, 7, 9$

III-3. **a.** $(y - 10)/72$ **b.** 18

III-5. **b.** 1/4

III-7. **a.** $b^2/2$

III-11. **a.** $a/4y$ **b.** $-a/4\ ln(a/4)$

Appendix B-V

V-1. **a.** Normal with mean $\dfrac{(\sigma^2\mu_0 + \sigma_0^2 x)}{\sigma_0^2 + \sigma^2}$ and variance $\dfrac{(\sigma^2\sigma_0^2)}{\sigma_0^2 + \sigma^2}$
b. $\dfrac{(\sigma^2\mu_0 + \sigma_0^2 x)}{\sigma_0^2 + \sigma^2}$

V-3. Gamma with parameters $m + x + 1$ and $1 + (m + 1)/\lambda_0$

V-5. **a.** 5.046 **b.** (4.87, 5.22)

Appendix VII

VII-1. **b.** $\hat{\sigma} = 0.082$ $\hat{\sigma}^2_{\tau\beta} = 0.013$
$\hat{\sigma}^2_{\beta} = 0.532$ $\hat{\sigma}^2_{\tau} = 0.362$

VII-3. **d.** $\hat{\sigma}^2 = 0.0596$ $\hat{\sigma}^2_{\tau\beta} = 0.0077$
$\hat{\sigma}^2_{\beta} = -0.0137 \to 0$

Index